9.95

Old Peninsula Days

Tales and Sketches of the
Door County Peninsula

With illustrations by
Vida Weborg

by

Hjalmar R. Holand

Wm Caxton Ltd
Sister Bay, Wisconsin

Published by

Wm Caxton Ltd
Box 709 — Smith Drive & Hwy 57
Sister Bay, WI 54234

(414) 854-2955

Printed in the United States of America.

10 9 8 7 6 5 4 3 2 1

Library of Congress Cataloging-in-Publication Data

Holand, Hjalmar Rued, 1872-1963.
  Old peninsula days : tales and sketches of the Door County peninsula
/ by Hjalmar R. Holand.
    p.   cm.
  Reprint. Originally published : Ephraim, Wis. : Pioneer Pub. Co., 1925.
  ISBN 0-940473-21-6 : 9.95
  1. Door County (Wis.) I. Title.
F587.D7H75   1990
977.5'63--dc20   90-2129
                                                                     CIP

ISBN# 0-940473-21-6 (paperback)

This book is printed on acid-neutral paper bound in sewn signatures and is
intended to provide a very long useful life.

PREFACE

Up to about fifty years ago every section of our land, surrounded by more or less definite physiographic boundaries, had a character of its own. It was marked by peculiarities of domestic handicraft, local traits and habits of speech.

But the railroad, the newspaper and the telephone have in most regions obliterated these diversities in population. The present generation has been molded in a form of trite uniformity, and individual resourcefulness has been superseded by the all-sufficient mail order house.

One of the last places in the West to yield to the demands of present day convention is the Door County Peninsula in northeastern Wisconsin. Sequestered as it is by large bodies of water on almost every side and far from the main lines of travel, its people have until recently retained an original freshness not often met with in this land of sophisticates. They have gone about their somewhat unusual daily occupations unembarrassed by a knowledge of outside standards and methods, and have thus developed types and a character of their own. Professor C. M. Moss of Urbana, Illinois, who for nearly forty years has spent his summers in Door County, recently wrote to the author, in the following words, of his early impressions of the place:

I cannot picture it as much less than a bit of the outer rim of Paradise. What with the rugged newness of the

iii

surroundings, its peaceful quiet so full of the very breath of serene life to one who could appreciate its soul invigorating influence, the simple, genuine kindness of the people, their calm and hopeful religious life, it resembled the sleep of a young child, trustful and undisturbed by the clamor and disquiet of the day, yet throbbing with a fresh life animating its being.

Northern Door County has also quite a unique history. The exceptional physical conditions which Nature has built up in and around the peninsula have produced men and experiences not commonly met with. The result is a pioneer history teeming with unusual characters and incidents.

This pioneer history is built on a background of Indian life and legend the counterparts of which may be present in other regions, but are there mostly forgotten. Here, happily, these ancient dramas have been preserved to us by the observations and remarks of the early French missionaries who visited and labored in this region more than two hundred years ago.

Recently this peninsula has become known to thousands of discriminating tourists attracted hither by its exquisite scenery and diversified flora. Here, far from the crowds and dust of the city, they have found a haven unharassed by the weary strain of social demands, and become almost like children again. "Here the wicked cease from troubling and the weary are at rest."

This book is written for the purpose of telling something about the making of this American community. The author has had the privilege of spending a quarter century of happy days in this fair region, and he knows it intimately. He has heard the story of its early settlement from the lips of the pioneers who conquered its wilderness; he has talked with scores of fishermen and sailors who have adventured on its waters; from the Indian and the pathfinder he has learned legends of old. From these sources and others has grown the following narrative, in which he attempts to picture the stirring traditions, the fascinating history, and the quaint characters that form the living background of this land of romance.

While this book is not intended as a local history, it is historical throughout. In two or three instances names have been changed, but otherwise it is true to fact. While the history of the peninsula abounds with many things of local interest, only such matters have been selected as are considered typical or have a common human appeal.

H. R. H.

Cedar Hill,
Ephraim, Wis.

CONTENTS

vii

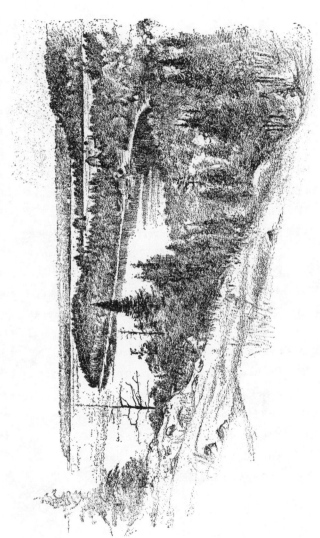

LITTLE SISTER BAY

CHAPTER I

PENINSULA PARK

Though all the bards of earth were dead
And all their music passed away,
What nature wishes should be said,
She'll find a rightful voice to say!
WILLIAM WINTER.

Far away, from the thousand hills of Wisconsin, the waters of Green Bay are gathered. They come purling out of gushing springs and gather into little rills rambling seaward. Into great rivers they join, like the swift Menominee, the somber Peshtigo and the famous Fox River. From cataract to cataract they leap, until at last they spread themselves on the bosom of the Emerald Sea.

Up at this head and beginning of the bay, the waters spread out sluggishly, with shallow, weed-grown bottoms, where the wild fowl stop to chatter on their flight to the tropics. Here the banks are low and the waters turgid, as if seeking rest. Perhaps they are weary of the busy hum of the cities that dot the Fox River valley, and fatigued turning the wheels of industry that line its banks. Gently the rippling waters roll seaward, caressing the timbered shores, tempering the chill western winds, and graciously giving to the peninsula that lies on its eastern border a climate of wonderful fitness for growing luscious fruits.

I

But the air of the Green Bay basin is invigorating, laden with life-stirring ozone from its evergreen forests. Soon, too, the sea feels new life pulsating within its deep. From the east to the west and on to the north it rushes, seeking a nearby opening to join its big brother Michigan.

Off to the east the green-topped hills fall apart, making a big opening for Sturgeon Bay. Into this bay the sea leaps gayly, believing it has found a channel to Lake Michigan. For eight miles the waves roll merrily inward, fed by this delusion, only to be stranded on the sands of a narrow isthmus, almost within sight of the big sea beyond.

Thwarted by this obstacle the sea now hastens out of the narrow confines of Sturgeon Bay and turns northward. It sweeps past the palisaded cliffs of Door County and hurls itself against the many crags and islands that oppose it. It throws big columns of spray and foam across Hat Island and Green Island and lashes the long-swept beaches of Chambers Island and the rocky strands of the Strawberry Islands.

At this place a rock-ribbed promontory, top-crowned with deep green woods, stretches far out across the path of the raging sea. Behind it rises two noble cliffs in perpendicular grandeur, leaning forward as if better to see what is disturbing the peace of this secluded region. For here is Peninsula Park, the sylvan dreamland of the North.

But the impatient sea cares nothing about sylvan seclusion. It pounds the limestone battlements of the promontory as if to grind it into powder. It thunders against the rock strewn shore with the sound of cannon. It booms and bellows in wild fury, throwing its white-flecked spray even to the tops of the trees. At last, its most violent efforts fruitlessly spent, it slips past the firm crags of Eagle Point, soon to be gathered up by the big gray brother it has been seeking.

But the waters around Peninsula Park are not always heaving in turbulent fury. On the contrary, that is exceptional. Most of the time the bay lies placid, gently caressing the pebbly beach, like a lazy kitten reaching out a playful paw. Smoother than glass and almost as transparent, the sea stretches out to a boundless horizon. Put a pane of glass into the bottom of your boat and you will see big tang-covered boulders strewn about. Among them grows an innumerable multitude of sea-bottom shrubbery, waving slightly as some strong finned denizen of the deep pushes his easy path among them. Here a school of minnows appears, a thousand or two, sporting in carefree opulence. Suddenly there is a wild scramble, as a rock bass shoots like lightning into their midst.

Nine miles it has of waterfront, varying with almost every boatlength in changing vistas. Here are craggy promontories with pinetops soughing in the wind. Here are sandy beaches, with a firm gravel bottom reaching out a quarter of a mile. Here are castellated precipices, almost attaining the dignity of mountains. Here are mysterious caves, opening to the breathing sea. Here are landlocked coves and reedy bayous, inviting to solitude and meditation, where the deep, romantic woods creep down to the shore, speaking to the traveler of a place where it is always afternoon.

But the charms of Peninsula Park are not all of the scintillating waterfront. Up high on top of those towering cliffs lies the Park proper, almost four thousand acres of hills and dales. Chicago has a generous park area, but all of Chicago's many parks could be tucked away here and yet leave room for a dozen good-sized farms. But no scissor-trimmed garden hedges here speak of man's puny efforts. Here Nature was the landscape gardener, mixing valleys and timbered slopes, open glades and lowly marshlands about in bounteous munificence. Here are glens, gleaming with coy wild flowers, hillsides glowing with blooming orchards, and dark forest recesses, where reigns the spirit of Pan.

Aye, the woods! The forest! Where, this side of Mariposa, can be seen such woods? All the trees of the North have here met and rule unitedly. Here is not only the sturdy oak, the graceful birch, the broad maple, the gorgeous cherry. Here are also the stalwart ash, the sinuous elm, the ancient beech, and the brawny ironwood. Here the brilliant sumac endeavors to outdazzle the glowing dogwood in color, hard pressed by the mountain ash and the luxuriant elderberry. The dainty locust, the mournful willow, the spreading butternut, and the stout basswood are here, and a score of others. Here, too, king among all, is the royal pine, while around him stand the worshipping balsam, the cheerful cedar, and the dreamy hemlock.

Where elsewhere can such a goodly company of ancient royalties be found? Where elsewhere can one walk beneath such noble temple arches? Strong and straight as the pillars in the temple of Karnak they stand. But they are not dead and decaying as those famous imitations of man. They are radiant with life and their heads a hundred feet up, bend gently and reminiscently to the salutations of the south wind.

Since the creation of the world we have ruled this ground, —they seem to say—and our crown shall overshadow it until Time's last crash of thunder is heard. The storms of centuries and milleniums have raged around us, the thunder of primeval times have pealed over our heads, and the earth has quaked under our feet. Yet, youthful as ever we stand. In the clefts of the rock have we fixed our feet, the snows of a thousand winters have we thrown off our limbs, and the summer floods have we sucked up as nourishment for our roots. The lightfooted deer have browsed on our underbrush in peace, the wolf and the bear have calmly nourished their young in our shade, and thy red brother has in days gone by gathered healing herbs between our feet. Now, white man, we would also be a solace and a benediction to thee. Take a lesson in serenity from us! Cease thy restless pursuit of idle phantoms, the lure of riches, the vain dictates of fashion, and frivolous gayety of the weakminded. Fulfill thy destiny as we have ours and thou wilt be a blessing to others as we have been to thee!

You, reader, who have sat in the shade of Eagle Cliff, towering behind you, "like a great rock in a weary land," with the bay before you like a silver field, your ears entertained by the

EAGLE CLIFF

murmur of the rockborn spring, your eyes delighted with curving shores and emerald islands,
while far up in the sublime heights an eagle is
majestically poised, you have felt that nowhere
is more perfect scenery. Other scenes may excel
in one thing or another, but nowhere is there a
more harmonious blending of all the elements of
scenic beauty. Is a Sicilian sunset more radiant
than that which meets the view from Sunset Cliff?
Are the waters of the Bay of Naples as limpid as
those which lave these shores? Is the Mediter-
ranean sky tinged with a fairer blue than that
which curves over Peninsula Park?

Door County has two hundred miles of sinuous
shorelines and much of this waterfront is as beautiful as that of Peninsula Park. Every craggy
promontory presents a most superb prospect, and
every indentation of the shore a pleasing pano-
rama. Its wooded hills with the blue waters of the
bay shimmering in the background are a balsam
to tired nerves, whether seen in the crisp green of
Spring or in the gorgeous coloring of Autumn
foliage. The perfect harmony of Nature speaks a
universal language, stimulating even the most
stolid to reflection and emotion.

It is because Peninsula Park is typical of this
land of beauty that a modest description of it is
here attempted.

CHAPTER II

IN THE DAYS OF THE INDIANS

In the vale of Tawasentha,
In the green and silent valley,
By the pleasant water-courses,
Round about the Indian village
Spread the meadows and the cornfields,
And beyond them stood the forest,
Green in Summer, white in Winter,
Ever sighing, ever singing.
                              LONGFELLOW.

Here and there, all over America, there are to be found scattered in the ground Indian relics which tell of the red man's presence long ago. These relics consist of implements of war, pottery, ceremonial implements, village sites indicated by kitchen middings and the remains of the arrow maker's workshops, grave mounds, cornfields, and pits for trapping wild animals. According to Indian archaeologists, Door County has a greater abundance of these Indian remains than any other region of Wisconsin. There are scores of village sites, some of them extending for miles, as on Washington Island and along Hein's Creek and Hibbard's Creek, north of Jacksonport. These extensive village sites are no doubt the sites of a series of villages occupied at different times. There are also very extensive cemeteries covering acres of ground, as on Detroit Island and on the extreme

8

end of the peninsula. These perhaps mark the site of great battles. There are all manner of other remains in such profusion that, as one archaeologist says, "there is little left to be desired."[1] All this shows that Door County in ancient days was a favorite abode for Indians. Indeed, Father Claude Dablon, who spent considerable time in the West about the close of the seventeenth century, says that this peninsula was "the Paradise of the Indians."

There is a good reason why the Indians found the Door County peninsula such a congenial place of habitation. Hunting was as good as elsewhere in the state, there was an abundance of maple trees for making sugar, and in addition to this they had the rich fishing in the surrounding waters.

La Potherie, a French historian who lived in the latter part of the seventeenth century, thus writes of the conditions of life among the Indians of the Green Bay region:

The country is a beautiful one, and they have fertile fields planted with Indian corn. Game is abundant at all seasons, and in winter they hunt bears and beavers. They hunt deer at all times, and they even catch wild fowl in nets. In autumn there is a prodigious abundance of ducks, both black and white, of excellent flavor, and the savages stretch nets in certain places where these fowl alight to feed upon the wild rice. Then advancing silently in their canoes, they draw them up alongside of the nets in which the birds have been caught. They also capture pigeons in their nets in the summer. They make in the woods wide paths in which they spread large nets in the shape of a bag and attached at each side to the trees;

[1] George R. Fox, in *Wisconsin Archaeologist*, January, 1915.

they make a little hut of branches in which they hide. When
the pigeons in their flight get within this open space, the
savages pull a small cord which is drawn through the edge of
the nets and thus capture sometimes five or six hundred birds
in one morning, especially in windy weather. All the year
round they fish for sturgeon, and for herring in the autumn;
and in winter they have fruits. . . . . This fishery suffices to
maintain large villages. They also gather wild rice and acorns.
Accordingly, the peoples of the bay can live in utmost
comfort.[2]

Amid such favorable conditions, why should not
great manhood and noble characters be developed?
True, the Indians did not attain many of the arts
and accomplishments of the present age, but
neither did the people of Homer. Yet, thanks to
his matchless pen, we know that among his people
there were some who were just as dauntless of
spirit, just as noble of mind, and just as lovable of
soul as are the people of today. Perhaps, if the
Indian had had a Homer, an interpreter, we would
find that among them, too, were men and women
of just as excellent moral fiber as Hector and And-
romache, just as sagacious as Ulysses, just as
steadfast as patient Penelope, just as pleasing as
the peaceful Phaeacians. They lived and loved and
suffered. They had their domestic and tribal
dramas. They had men of great personality rise
in their midst and draw the admiration of far and
near. But of this we know little or nothing. And
knowing nothing, we simply think of the Indian as
an indolent and blood thirsty brute, without a
soul, a cross between an animal and a fiend!

[2] La Potherie's *Amer. Septentrionale*, pp. 80, 81.

Human nature is much the same in every age and in every community. Given the poet, the true historian, who can rightly interpret the achievements of his people, these would in almost every case attain almost epic grandeur.

There is a side to the Indian character of which we know little. We see it dimly typified through the fog of oblivion in the person of Tomah, the great chief of the Menominees and overlord of all the Indians of the Green Bay region, the Pontiac of Wisconsin, who could boast that his hands had never been defiled by human blood. It is also typified in the creation of the Hiawatha legend which, according to Schoolcraft, had its birth in the same region. The creation of such characters as the industrious Hiawatha, the wise Nokomis, and the gentle Minnehaha, show us that they were not foreign to the ideals of the people that adopted them.

But the peaceful tradition of this side of the Indian's nature have been overshadowed and forgotten by the more dramatic recollections of the great wars they took part in. The traditions that have been preserved nearly all deal with the great wars, the terrible sieges, the awful pestilences that the Indians of the peninsula suffered in the latter half of the seventeenth century, and which made such a profound impression upon their minds. The Green Bay region, and particularly the Door County peninsula, were for ages the battleground, first between the Algonquins and the Sioux, and then between the Algonquins and the Iroquois.

When the various Algonquin tribes, the Ottawas, the Sacs, the Foxes, the Potawatomis, the Menominees, were driven westward by the warlike Iroquois, they found this new country even more pleasant than the lands from which they had been expelled. Here they were content to dwell in a land of plenty, but here too they were opposed by the terrible Winnebagos, who had come from the southwest and subjugated all the country before them. Between these two tribes, the Algonquins from the East and the Winnebagos, a branch of the Sioux, from the Southwest, there followed a long period of warfare of varying success (with interludes of savage attacks by the Iroquois from Lake Ontario), which seems to have reached its climax during the middle decenniums of the seventeenth century.

Just at this time and immediately afterward it happened that the missionary enterprises of the Jesuits in the Green Bay region were undertaken. These missionaries sent to their superiors in Quebec reports of everything pertaining to their work. By help of their fragmentary and unadorned records we have the curtain raised for a few years on a series of stirring dramas, at a time when the passions of the Indians of this region were at their highest. We see them toil, scheme, fight, suffer, and die, no doubt much in the same way as their ancestors from time immemorial had done.

When Nicolet arrived, he found two groups of Indians on the Door County peninsula. One was a village of mixed Indians at the mouth of the bay, the other was the Winnebagos, strongly intrenched and fortified in a community containing four or five thousand inhabitants, at a place called Red Banks, twelve miles northeast of the present city of Green Bay.[3] Of the village at the mouth of the bay, we are told that it was "composed of people gathered from various nations (Potawatomis, Sacs and Menominees), who, wishing to commend themselves to their neighbors, have cleared some land there, and affect to entertain all who pass that way. Liberality is a characteristic greatly admired among the savages; and it is the proper thing for the chiefs to lavish all their possessions, if they desire to be esteemed. Accordingly, they have exerted themselves to receive strangers hospitably, who find among them whatever provisions are in season. And they like nothing better than to hear that others are praising their generosity."[4]

This village of kindly disposed Indians "at the entrance to the bay" was no doubt on Washington Island, where there are numerous village sites, showing that it has been populously inhabited by Indians. It is claimed by archaeologists that no tract of equal area in the state of Wisconsin exhibits so many evidences of Indian occupation as Washington Island.[5]

This village was probably made up chiefly

[3] *Jesuit Relations* (Cleveland reissue), XXIII, pp. 275-279.
[4] La Potherie, page 69.
[5] See Geo. R. Fox's article on *Indian Remains on Washington Island*, in the *Wisconsin Archaeologist*, Madison, Wis., for January, 1915.

of Potawatomis, as these islands from early times
have been called Potawatomi Islands. It is evident
that Nicolet stopped among these hospitable and
intelligent Indians for some time and from them
obtained the information about the Indian tribes
of the region which was later recorded in the
"Jesuit Relations" and by La Potherie.

The Winnebagos, to whom he was journeying,
were at that time the most formidable people of
Wisconsin. Accordingly to Jonathan Carver, who
spent some time with them in 1766,[6] they originally
came from Mexico, and had traditions of battles
with the Spaniards, whom they intensely hated.
They were the only tribe in Wisconsin that used
horses and they were the fiercest of warriors. They
were very perfidious, superstitious, and insolent,
and not only took the scalps of their enemies, but
devoured their bodies at their feasts. When they
came to the Door County peninsula, they found
the land occupied by the Potawatomis. The latter
are described as being the most affable and
generous of the Wisconsin tribes, and were willing
to divide the land with them, as there was plenty
for both. But the arrogant Winnebagos would
listen to no peace proposals, and made war upon
the scattered villages of the Potawatomis at every
opportunity. Fleeing before superior numbers, the
Potawatomis were finally compelled to leave the
peninsula and take refuge among their brethren
upon the islands, where was also a village of

[6] See Carver, *Travels through the Indian Parts of North
America in 1766-1768.*

Menominees and Sacs. Yet not even here would the insatiable Winnebagos leave them in peace. They marched an army to the north end of the peninsula, made canoes, and with human sacrifices and invocations of the Great Spirit made ready for an onslaught upon the islanders.

The Potawatomis were early apprised of the coming of their enemies. After hurried deliberation, the chiefs decided that their best course was to send an army across the strait and attack the Winnebagos in the rear at such a time as they might be unprepared for attack. Three spies were therefore sent across with instructions at the right time to build a signal fire upon a certain bluff, by which the Potawatomis would be guided in making a landing on the rocky and dangerous shore.

These spies unfortunately fell into the hands of the Winnebagos who subjected them to frightful tortures. Rather than reveal their mission, two of them perished at the stake. The third, however, was finally bribed to disclose his secret.

With great glee the crafty Winnebagos now prepared to turn this stratagem to the destruction of their enemies. Upon a dark and windy night, the signal fire was built, not upon the bluff selected by the Potawatomis, but upon another nearby, whose precipitous base afforded no landing place. Simultaneously they sent a small detachment of warriors in canoes by a circuitous route to fall upon the defenceless camp of the Potawatomis.

Encamped upon Detroit Island, their canoes ready, the Potawatomis saw the appointed signal

THE END OF THE PENINSULA

fire leap into the air. Several hundred strong they immediately pushed off, regardless of the wind which was not favorable for crossing. During the passage the wind increased with violence, so that they could not have returned if they had wanted to. However, they made the crossing in safety, but instead of finding a beach favorable for landing, they found their frail canoes thrown against precipitous rocks upon which they were crushed like eggshells. Some attempted to turn back, but their canoes were quickly swamped in the breakers. Others clung to the crags and roots of trees, but were quickly tomahawked by the exultant Winnebagos.

A shelf of rock jutted out at the base of the cliff just out of reach of the waves. Here about thirty of the Potawatomis managed to clamber up with their tomahawks in their belts. Standing here between the roaring sea, and the fierce Winnebagos above them, they chanted their death song, defying the Winnebagos to come and get their scalps. With a thirst for blood like wolves, the younger men among the Winnebagos could no longer contain themselves, but leaped down upon their prey. Here they were met by the Potawatomis, who cleaved them with their tomahawks as they fell. Again the eager Winnebagos leaped down in greater numbers than before. Each seized his enemy and a desperate struggle ensued, when suddenly a great wave came out of the stormy deep and sucked the fighting savages off the shelf into the sea where they all perished.

The canoe party of the Winnebagos fared no
better. Soon they found themselves wallowing in
a terrible sea out of sight of land. Unused to
navigating the rough waters of this channel, their
canoes one by one were swamped, and they all
perished.

For a whole day the Winnebagos stood upon the
cliffs watching for the return of their victorious
canoe party. But none returned. When they saw
the wreckage of their canoes drift in on the shores,
they understood that their brethren had perished
in the storm. They took this as an omen that they
must never attempt to cross that "Door of Death,"
which it was afterward called.[7]

[7] This tradition was received from the Indians who lived
there by the early fishermen who settled on the islands about
1840. It is also mentioned by several of the early travelers.
(See Storrow and Stambaugh.) Captain Brink, one of the
government engineers who surveyed this part of Door County
in 1834, mentions this tradition as follows: "The Indians say
that a whole tribe of Indians, three hundred in number, lost
their lives one night near the big bluff that is called 'Death's
Door,' and that is why the spot was given such a dismal
name. As the story goes, the tribe was to land in canoes near
the spot where they are said to have lost their lives, and surprise their enemies, who were encamped near by. They were
betrayed, however, by one of their number who was to notify
them on the beach as to the best place to make the landing.
Instead of building the fire on a hill about a quarter of a mile
further up the lake, the signal was placed on the bluff, and
when the three hundred Indians attempted to make the
landing they were dashed against rocks and perished. That
is how the spot is said to have obtained its name." The
French early adopted the Indian name of the channel and
translated it into Porte des Morts, Death's Door. This was

This battle almost wiped out the entire force of Potawatomi warriors. Fearing another attack by the Winnebagos, the survivors fled from the islands, taking up their abode on the other side of Green Bay for the time being. Meanwhile their arrogant conquerors believed themselves the most powerful people on the earth. Their villages were scattered all the way down the peninsula,[8] and from here they meditated expeditions of conquest against more distant nations.

A few years after these events (about 1650), a war was waged between the Iroquois and the Ottawas (including the Hurons) in the country southeast of Lake Huron. When the latter were overcome they fled westward. Finding Washington Island deserted they settled there. Still fearing attacks the Ottawas kept a scouting party of picked men on foot in their old country near the present city of Detroit, to give them notice of the plans of their enemy.

After some time these scouts learned of a force

---

later abbreviated to "the Door." As the first settlers of Door County at the time of its organization lived down at "the Door," this was taken as the name of the county.

[8] Spoon Decorah's recital of the traditions of his people in *W. H. S. Coll.* XIII, page 457, says that before the Winnebagos moved to Red Banks, they lived below (north of) Red Banks. "There was a high bluff there that enclosed a lake." This is, no doubt, a reminiscence of the precipitous shore of northern Door County, from which Green Bay looks like a lake. La Potherie, who got most of his information from Perrot, also calls Green Bay a lake and says that the Winnebagos were masters of it and a great extent of the adjoining country. *Histoire de l'Amerique Septentrionale* (1722) p. 70.

of eight hundred Iroquois who were bent on further destruction of the Ottawas and were seeking their place of refuge. They hastened back to their tribe and acquainted the chiefs of the enemy's approach.

As Washington Island did not lend itself to a strong defense because of the lack of running water, the Ottawas and Hurons moved southward along Lake Michigan. After a time they found a stream of water with open sandy fields suitable for growing corn. On both sides of the stream the men built a palisaded village, while the women planted fields of corn. In both of these under-takings they were successful, and, before the Iroquois had discovered their refuge, they had time to finish their stockade, harvest their corn, and bring in a large amount of game. Topo-graphical conditions limit the location of this Indian fort to Hibbard's Creek, a short distance north of the village of Jacksonport.

Finally one day the Iroquois discovered their stronghold and with savage yells made a furious onslaught on the stockade. But this was built of green logs thirty feet high, standing close together and firmly buttressed on the inside. In vain they tried to scale it, to hew it down, to burn it. All these attempts only brought death to the fool-hardy attackers.

Seeing that all such attacks were in vain, the Iroquois now settled down to besiege the village and compel the Ottawas to yield by starvation. In the meantime, however, the Iroquois had great difficulty in feeding themselves. The Ottawa

hunters had cleaned the vicinity of game and fishing was not always successful. Meanwhile the Ottawas lived in abundance and, like the ancient Roman general, threw loaves of bread over the stockade upon the heads of their besiegers to show how futile their attempts were.

The Iroquois, after a time, realized that their plans of starving the Ottawas into surrender were doomed to disappointment. They also became aware of the dismal fact that they themselves were in danger of starvation. Finally they were obliged to make humiliating. terms of peace, whereby they were to buy food supplies from the Ottawas and thereupon depart.

The Ottawas, rankling with revengeful feelings for the injuries they had suffered from the Iroquois in the past, now planned to teach them a lesson. On the day before the departure of the besiegers, the Ottawas baked a great quantity of corn bread, into which they mixed a deadly poison. They then announced to the Iroquois that, in token of their friendship, they desired to present each Iroquois warrior with a loaf of bread. This announcement was greeted with great joy by the famishing Iroquois.

The mother of a certain Huron warrior among the Ottawas was a slave in the camp of the Iroquois. Fearful of her safety, this Huron told his mother in great secrecy not to taste of the bread, as it contained death. She was led to reveal this secret to her master, the chief. When the bread was tossed down to them, this chief gave part of his loaf to a dog which, upon eating it, died in great agony. Gloomily the hungry Iroquois departed, followed by the jeers of the Ottawas.[9]

[9] In Perrot's *Memoire sur le moeurs, coustumes et relligion des sauvages de l'Amerique Septentrionale* (pp. 80-83) where this tradition is recorded it is stated that the Ottawas and Hurons settled on Huron Island. This has led some commentators to believe that this place of refuge was Huron Island, in Lake Superior, some distance northwest of the city of Marquette. This conclusion is impossible, as Huron Island is very small, entirely too small to have been chosen as the home of two tribes counting hundreds of warriors, and their families. Jules Tailhain, the learned editor of Perrot's *Memoires*, thinks that it was Washington Island. This is no doubt right, because we learn from other sources that the Ottawas inhabited this particular region at the time of the Iroquois expedition to the West which, according to Perrot, took place in 1653-55 (see op. cit.). The Ottawas are not mentioned among the tribes of this region until about the time of the overthrow of the Winnebagos at Red Banks, which, according to Father Allouez, took place about 1650 (see following pages). When the Winnebagos fled from Red Banks their stronghold there was occupied by the Sacs who remained for some years (about 1655-60). After a few years these Sacs were overthrown and almost exterminated by a coalition of the tribes of the Bay, consisting of Menominees, Chippewas, Potawatomis and Ottawas (see following pages), in which the Ottawas are said to have taken the leading part. (Cf. the Chippewa tradition of this battle recorded in *Wis. His. Coll.* XV, pp. 448-51, and the Menominee tradition in *Ibid.* II, pp. 491-94.) As the detailed accounts of the further wanderings of the Ottawas show them to have spent the remaining years of that century in western Wisconsin, at Chequamegon Bay on Lake Superior and on Manitoulin Island in Lake Huron, this limits their stay in northern Door County to the years immediately before and after the Iroquois incursion.

The Ottawas, fearing a renewed attack in larger force from their implacable enemies, thought it wise policy to ally themselves with their formidable neighbors, the Winnebagos. They therefore sent envoys to them, bearing messages of peace and friendship and many presents, such as knives, needles, hatchets and articles of ornament. The insolent Winnebagos, however, disdainful of any alliance, received the messages and presents, and thereupon killed and ate the emissaries.

This shocking outrage greatly incensed not only the Ottawas, but all the other tribes in the Green Bay region. A league of war was formed to ex-

Other evidence which points in the same direction is as follows: (a) When the Ottawas and Hurons left the island they are said to have "retreated to Michigan," which can only mean that they established themselves at some point on Lake Michigan. (b) When the Iroquois withdrew from their futile siege they divided their forces, one part "pushing further on [up the lake], until they encountered the Illinois," while the others fell in with the Chippewas. Both of these tribes lived on the lands adjoining Lake Michigan, the one at the south end, the other at the north. (c) When the Ottawas withdrew from their fortified village after having repulsed the Iroquois, they are said by Perrot to have gone up the Fox River and down the Wisconsin. This shows that the battle ground was north of the Fox River. (d) The only spot on Lake Michigan north of the Fox River having a small perennial stream of water (necessary in a besieged village), is Hibbard's Creek, a short distance north of Jacksonport in Door County. Here very extensive Indian remains have been found for a distance of three miles up the creek. The land on both sides of the creek is sandy and suitable for the Indian method of cultivating corn.

terminate the Winnebagos, and many expeditions
were made against them. For a while the latter
held their own, but finding their scattered villages
were exposed to surprise attacks by the enemy,
they moved southward and concentrated at Red
Banks, a high hill overlooking Green Bay. Here
they built a great stockade, with a deep ditch or
moat on the outside. Still "numbering four or five
thousand men" they were able to resist their
enemies successfully.

Before long, however, another enemy more
terrible than the Ottawas overtook them. The
filth caused by their congested quarters resulted in
a great sickness which killed thousands of them.
For a while the terrible plague swept through the
great town so rapidly that they were unable to
bury the dead, who lay rotting in the sun, while
the survivors lay upon their beds of sickness,
groaning with pain. Within a short time their
number was reduced to fifteen hundred men.

Yet even this dreadful visitation did not stop
their plans for conquest. No sooner did the plague
pass away, than they determined to send a strong
force against the Outagamis, or Foxes, a numerous
people who lived on the other side of the bay. This
was probably about where the present village of
Pensaukee is located. Five hundred warriors
fitted out with arms and warpaint set out in
canoes to cross Green Bay. On the way a gale of
wind overtook them and the entire force perished.

Finally, to fill their cup of misery, the Winnebagos were threatened with starvation. The many

raids which their enemies had made upon them had dispersed the game, and now it came to pass that it was impossible for them to find the necessary food.

Upon seeing all these afflictions overtake them, their old enemies, the Potawatomis, the Ottawas, and other tribes, had compassion upon them and forbore to attack them any more. One tribe even went further. The Illinois, a noble minded people, living some distance away, took pity upon the famished Winnebagos and sent five hundred young men to them with food and presents. Among these five hundred were fifty chiefs of the Illinois.[10]

But all the adversities that had overtaken the Winnebagos had in no wise changed their perfidious hearts. Their medicine men asserted that the souls of their departed, especially those slain in battle, could not rest in peace unless their relatives avenged their death. The five hundred men sent to fight the Outagamis had met their death, and not one enemy had been slain to avenge them. They therefore longed with superstitious zeal to find some victims whose blood would relieve the shades of their own departed from further unrest.

When the news came of the approach of the five hundred Illinois, laden with gifts, the medicine

[10] In Saint-Lusson's *Proces-Verbal*, June 14, 1671, quoted in *Wis. His. Coll.* XI, pp. 26, 27, the Illinois are twice mentioned as being neighbors of the Ottawas, Potawatomis, and Sacs. At the time of these events they probably lived in southern Wisconsin.

men worked themselves into a religious frenzy.
Here, they declared, was just the right number of
victims to atone for the death of their own warriors. The Illinois messengers must be seized and
sacrificed to the shades of the departed or the
plague would return and devour the remainder of
the people. Somewhat perplexed about the pru-
dence of this course, the people finally adopted it.

The Illinois arrived and they and their gifts were
received with a great show of gratitude and rejoicing. A huge cabin was erected to house the
guests and sports were arranged for their entertainment. Then the Winnebagos made ready a
dance for their guests. While the unsuspecting
Illinois became more and more animated in the
dance, the treacherous hosts cut their bow strings
and then, upon a signal, flung themselves upon
the dancers. All were massacred, not one escaping,
and a great feast was made of their flesh.

When this horrible orgy was over, even the
perfidious Winnebagos were filled with remorse
over their treachery and ingratitude. The Illinois
were a mighty nation and they knew that sure
punishment would be meted out to them. They
did not dare to remain in their fortress at Red
Banks, but moved to an island in Green Bay,
where they thought they were safe because the
Illinois did not use canoes. But the Illinois
learned their whereabouts and waited until the
bay was frozen over. Then they marched to the
island on the ice, but found that the whole tribe
of Winnebagos, now greatly reduced by the plague

and starvation, had left the island in a body on a hunting expedition. The Illinois pursued them and after six days caught up with them. In the resulting battle, all of the Winnebagos except one were killed or taken captive.[11]

The old Winnebago capital at Red Banks was now occupied by the wandering Sacs, who waxed very mighty and arrogant in their new abode. They were a brutal, savage tribe who lacked the common courtesies of the Indian people and soon gave offense to all. They frequently made prisoners of peaceable Indians passing that way, and sacrificed them upon a scaffold elevated above the ground for that purpose. Before the prisoner was burnt, he was required to name the principal men of his tribe, whereupon these were reviled and burnt in effigy.

At one time an Ottawa Indian and his wife were taken prisoners and preparations were made to burn them. When the Ottawa was placed upon the scaffold, his courage failed him and he screamed with fear. At this the Sacs howled with joy. His wife then sprang upon the scaffold and said: "Your unmanly weakness gives pleasure to these devils. Let me show you how to die!" Thereupon

[11] For the above facts, see La Potherie's *Amerique Septentrionale*, pp. 69-81. Father Claude Allouez, who for years labored among these tribes, corroborated this story and adds: "About thirty years ago (from A.D. 1677), all the people of this nation (the Winnebagos) were killed or taken captive by the Illinois except one man." Jesuit Relations, LIV, p. 237. See also Charlevoix, *Journal Historique*, (Paris, 1744), pp. 290-296.

she was tied to the stake and asked to name the chiefs of her tribe. Defiantly she told them that she would give them the names of no worthy men to insult, but that she would name one Sac chief who would soon meet his doom because she had six brothers who would avenge her. With this she pointed to the principal Sac chief and was burnt with a song of defiance upon her lips.

Little by little this story found its way until it reached the ears of Nangadook, her oldest brother, at the mouth of the bay. In silence Nangadook heard of the brave death of his sister. Then he sent the warbelt and war pipe to all the tribes in the north country, and asked them to join him in punishing these insolent Sacs, who had treated with insult all the chiefs of the tribes that dwelt in that region. They came, the Potawatomis, the Menominees, the Chippewas, and many others, a flotilla of canoes filled with warriors. When all had assembled they proceeded up the bay to Red River, a few miles north of the Sac village. Here their canoes lined the beach for two miles so thick that no more could be crowded in. Silently the army made its way through the forest in the darkness of the night until they had completely encircled the village of the Sacs. Then before proceeding to attack, they laid down to rest.

A young Sac woman had that same night been given against her will to a Sac brave who lived some distance from the village. During the night she ran away from her husband and returned to her father. On her way, she passed the line of

sleeping besiegers and discovered that an unbroken line of an unknown enemy encircled the
village. With terror she ran to her father's tepee
and awakened him saying, "We are all dead!"
However, thinking only of how her husband had
dragged her off the night before, her father only
laughed at her and reproached her for her temper.
Enraged by the ill treatment she had received
both from her husband and father, she said no
more but rolled up in her blanket and went to
sleep.

The Sacs were awakened early the next morning
by the battle cry of their allied enemies and a
terrible onslaught ensued. A part of the stockade
was broken down, and the Sacs, taken unaware,
were slaughtered in great numbers. Yet they
rallied, drove the invaders back, and fought stub-
bornly all day, and many were killed on both sides.
For many days and nights afterward the battle
raged with great fierceness, but the deep moat dug
by the Winnebagos and the high wall protected
the Sacs. But they were soon attacked by another
enemy, thirst. Again and again they made
desperate sallies down the steep slope to get water,
but were each time met by the watchful besiegers.
The slope was soon strewn with the dead.

The heat of the burning sun and the terrible
suffering from lack of water, became intolerable
to the Sacs, and they saw that they would soon
miserably perish from thirst, even if they could
repel the attackers. So they planned a stratagem
to escape. Standing one evening upon the walls,

waving a flag of truce, their heralds called to the
besieging chiefs and said: "Our great chief,
Akheenandodaug, is dead. Give us tonight to
bury him in a manner befitting his rank and we
will resume fighting tomorrow!" This prayer was
granted and the besiegers, weary with much
fighting, slept heavily that night. The next morning when they prepared to resume the assault, they
were amazed to find that the enemy was not there.
Stealing quietly out between the sleeping ranks of
the enemy, the Sacs had fled far to the southward,
not stopping until they reached the upper Fox
River, above Lake Winnebago. After pursuing
the enemy, Nangadook returned with his forces
and leveled the old Sac and Winnebago capital
to the ground. Since then no Indians have had a
village on that cursed spot.[12]

After the defeat of the Sacs, no great battle or
feud has taken place among the Indians of the
peninsula. A band of Ottawas settled on Detroit

[12] The above narrative of the Sacs at Red Banks is handed
down to us in two independent traditions, one from the
Menominees (as narrated by Charles D. Robinson in *Legend
of the Red Banks*, in *Wis. His. Coll.*, II, pp. 491-494), and the
other from the Chippewas (as narrated by George Johnston
in *Osawgenong—A Sac Tradition*, in *Wis. His. Coll.*, XV,
pp. 448-451). Although they differ in minor details, there can
be no doubt that they tell the same story. The parallel walls
twenty-five feet long in the centre of the village, which
Mr. Robinson saw in 1856, were probably the base of the
sacrificial altar mentioned in the narrative. Mr. George Fox
also heard the tradition in the old Indian haunts on the Pensaukee, Little Suamico, and Big Suamico, where was the
great village of Oussouamigong. See his article in *Wisconsin
Archaeologist*, April, 1913, pp. 127-129.

Island, which came to be looked upon as the ancient seat of their tribe.[13] The Potawatomis after a time moved away from the islands and resumed their peaceful dominion of the peninsula. The many war parties which had overrun the peninsula had greatly depleted the game and it did not offer the happy conditions of life as formerly. When therefore Captain Cadillac a few years later invited them and other tribes to settle in southern Michigan to build up a great trading post at Detroit and to close the way against the Iroquois, the Potawatomis accepted his invitation and moved to southern Michigan. Here they built a village at the mouth of the St. Joseph River and another at Detroit. Only a few remained on the Door County peninsula. About 1812 some of the Potawatomis from St. Joseph River moved to Chicago, where they perpetrated the Fort Dearborn Massacre.

As to the terrible Winnebagos, that Ishmael among Indians who once played such a savage part on the peninsula, their power was thoroughly broken. When the Illinois released their captives, a few returned to their old hunting grounds, where they were looked upon as thieves and robbers. The larger number settled on the shores of Lake Winnebago. From here they spread westward, settling at the headwaters of the Fox River in the vicinity of the present city of Portage, where Jonathan Carver and Peter Pond found them in the latter part of the eighteenth century. The

[13] See Samuel A. Storrow's *Northwest in 1817*, in *Wis. His. Coll.*, VI, p. 165.

first settlers who came to Wisconsin, about 1830, found them in southwestern Wisconsin and northwestern Illinois, claiming possession of the lead mines, and in northeastern Iowa. Because they were in the way of the immigrants who were coming westward, in 1848 they were deported four hundred miles into the wilderness, to the present Todd County, Minnesota. They refused to stay there and many came back to Wisconsin the following year. They were then taken to the vicinity of Mankato, and later, far up the Missouri, to the Crow and Creek reservation. The government was unable to keep them there, and a reservation of 128,000 acres was set aside for them in Dakota County, Nebraska. By 1873 most of them were back in Wisconsin again, whereupon another order was issued to deport them. They were kicked and beaten and handcuffed and prodded with bayonets, but still clung to their native soil. In spite of their hardships and many deportations, there are still about fifteen hundred left in Wisconsin, living mostly north of Tomah and east of Black River Falls, where they receive an annuity from the Federal Government.

The above account is called "traditions," which does not give a correct impression of its significance. It would be more correct to call it history in practically all respects except definite dates and details. It is corroborated by, and largely based upon, the reports of the early Jesuit missionaries and fur traders who sojourned in this region immediately after the events recorded took place.

CHAPTER III

We were the first that ever burst
Into that silent sea.
COLERIDGE.

The French dominion in America may be said to begin with the year of 1608. In that year Samuel de Champlain made a settlement at Quebec. He explored the coasts carefully, penetrated up the great St. Lawrence River with its many tributaries, and was the first to behold the great chain of inland oceans that stretched for unknown distances to the westward.

Champlain's chief object in establishing a colony in the new world was to develop the fur trade. Prowling through the forests of this new continent were tens of thousands of Indians. They lived chiefly by hunting and had peltries by the millions. These were of little use to the natives, but of great commercial value to the French. Champlain's first object was therefore to gain their good will and their trade by a conciliatory and fraternal attitude toward the red men.

When Champlain established his colony at Quebec he found the vast wilderness around him dominated by two groups of Indians. These were the Iroquois and the Algonquins. Roughly speaking, the Iroquois ruled over the region south of the St. Lawrence while the Algonquins occupied the

33

northern country. Champlain soon found that the
two groups were irreconciliable enemies and he
had to choose between them for his field of enterprise. He chose the Algonquins.

When he had reached this decision, he took
steps to cement his alliance with the Indians in a
very sagacious and far reaching manner. He endeavored to teach them the fear of God by the
establishment of missionary stations, and to
protect them against their ancient enemies, the
Iroquois, by the establishment of a series of frontier
forts which became havens of refuge in the inter-
mittent wars between the two Indian confedera-
tions. He worked out his plans with excellent
results. Henceforth we see the quaint alliance of
missionary and merchant, the black-robed Jesuit
and the dealer in peltries. So successful was the
policy of the French in dealing with the Indians,
that as long as the French continued in power in
Canada, their Indian allies never turned against
them. While the English cavaliers on the St.
James and the English Puritans in Massachusetts
were in constant dread of extermination, the
French in Canada ruled peacefully and profitably,
their name honored and respected for a thousand
miles westward.

Little by little Champlain extended the sway of
the French dominion westward by making treaties
with distant tribes and establishing friendly
relations. In this work his chief agent was a man
by the name of Jean Nicolet. He came to Canada
in 1618 and soon showed a remarkable aptitude

for getting along with the Indians. He was therefore commanded by his superior to go and live
among the various Indian tribes, to learn their
languages, their manners and customs, their
secret thoughts. This was done. Nicolet for years
lived among the Indians and became as one of
them. He taught them many clever arts and
accomplishments of the white man, and in turn
learned like an Indian to bear suffering without a
murmur and to go days without food. So great was
his endurance, his tact, his ability, that he won the
confidence of the Indians, and was adopted into
many tribes, made chief, and was looked upon as
the most wonderful man that ever came to their
knowledge. By his cleverness and tact he was
able to make treaties of peace and friendship with
tribe after tribe, preparing the way for the missionary and the fur trader. He even accomplished
the unique feat of persuading the relentless Iroquois to bury the hatchet of war and make a
treaty of peace with their traditional enemy, the
Algonquins.

Largely through the power of his persuasive
personality, the fame of the great French Father
at Quebec spread far and wide and Indian
envoys came from unknown regions to pay their
respects. Among them came one summer day
a naked Chippewa, from a distant country "two
moons' journey" to the west, and laid at Champlain's feet a lump of copper.

On being questioned he told of the long journey
he had made and of the Great Lakes, the inland

oceans that stretched for unknown distances to the west and southwest of his native country. He told of the Indian tribes that lived about the shores of these lakes, one of which particularly interested Champlain. These were the Winnebagos, or, as their name means, the fetid or salt water people, the most savage and crafty Indians of the West and the chief tribe of that region.

Like all other explorers of his time, Champlain had the same hope as that which inspired Columbus, that of finding a passage to Cathay, or the Orient, the land of spices, pearls, and untold wealth. When he heard of these great seas with the name of the people living upon them suggesting that they lived in proximity to the ocean, he thought that they were probably one of the outlying tribes of China. To ascertain this and to extend the French empire to this distant and powerful people, he decided at once to dispatch Nicolet to negotiate a treaty of peace with them.

This journey of Nicolet's, a thousand miles into the interior of an unknown wilderness, inhabited with ferocious warriors, is one of the most fascinating events of American history. Yet never was a great journey undertaken with less flourish of trumpets. Fifty years ago Henry Stanley made his sensational journey of seven hundred miles into the interior of Africa to look for Livingstone, and the world has not yet ceased to wonder at the bravery of it. But Stanley had an army of camp followers and assistants, camels and beasts of burden, plenty of arms and a ponderous luggage.

EAGLE POINT SHOWING EAGLE ISLAND TO THE RIGHT

Nicolet had none of these things. Alone and almost
unarmed he seated himself in a birchbark canoe,
taking only seven dusky Hurons to use the
paddles for him.

As he started out upon this great journey we
may well imagine some of the woeful prophecies
that were made to him; of the turbulent rivers
and fearful rapids he would have to breast; of
the huge seas on which raged such terrible storms;
of whirlpools, of supernatural dangers, of blood-
thirsty Indian tribes who knew no mercy, but
killed and ate all strangers who came to their
shores.

But Nicolet was not dismayed. He was thoroughly familiar with the Algonquin language and
perfectly understood the innermost recesses of
the Indian mind. Pushing up the pine clad gorges
of the turbulent Ottawa River with its hundred
waterfalls, he reached Lake Nepissing, far up in
the north country. From here a river led to Geor-
gian Bay. Then followed a journey of many hundred miles along the shores of Lakes Huron and
Michigan. Finally after about ten weeks of
canoeing, he reached the shore of the Door County
peninsula.

On this long journey Nicolet visited many
Indian tribes who had never seen a white man.
Some of these savages hailed him as a god, while
others fled from him in terror. But Nicolet's
commanding presence and reassuring smile were
able to overcome all their distrust and fear.
Although their training and experience led them

to look upon strangers with suspicion and hostility, he was able to wander at will among them, honored and unmolested.

The Winnebagos, "Men of the Sea," toward whom he was traveling were at that time occupying the greater part of the Door County peninsula. Their main stronghold was a large, palisaded village at Red Banks about twelve miles northeast of the present city of Green Bay. When Nicolet was within two days' journey of this village, he stopped and sent one of his Huron companions to them with greetings of peace and to apprise them of his coming, according to Indian custom. This camp site of Nicolet was probably on Eagle Island, near Ephraim, which was the favorite camping place because of its excellent harbor. It also agrees with the tradition of the Menominee Indians that Nicolet was met by the Winnebagos at his camp twenty miles north of the mouth of Sturgeon Bay.[1]

When he arrived at the Winnebago village, Nicolet took it for granted that he had reached an outlying settlement of the Chinese of the Orient.

[1] It has been supposed by some eastern writers that Nicolet's camp was on the west shore of Green Bay. There is no evidence whatever in support of this view. The west shore was seldom frequented by travelers in boats because it has a dangerous surf and few good harbors. Moreover, as the Winnebago village lay on the east shore, there would be no object for Nicolet to make two long and unnecessary trips across this large bay. Finally, if Nicolet's camp had been on the west shore, the Menominees, who then lived on the west shore, would not have a tradition that his camp was on the east shore.

He therefore arrayed himself in a highly decorated
Chinese robe brought for the occasion, and, firing
off pistols in the air, entered the village with such
pomp as his meager resources permitted. The
Indians were greatly impressed, having never
heard firearms before. The women and children
fled, while the warriors made humble salutations,
believing that the great Manitou had deigned to
visit them. He was therefore lavishly entertained
and treated most respectfully.

A short time after Nicolet returned from his
western trip, Champlain, the enterprising governor
of the French possessions in America, died. With
him also passed away the spirit of exploration for
many years. His successor, Montmagny, was not
interested in extending the French dominion to
remote regions, but quietly governed the affairs
of the colony at home.

RADISSON AND GROSSELLIERS

About twenty-five years after Nicolet's visit
we come to the next visitors. These were Radisson
and Grosselliers, two picturesque and romantic
soldiers of fortune, who traveled about for many
years in this savage wilderness, apparently with
no greater object than to seek adventure and "to
be known with the remotest peoples." In both of
these ambitions they were abundantly gratified.
They stayed for a while with an Indian tribe,
joined them in their wars, performed wonders of
slaughter with their guns, and then went on to the
next tribe to be worshipped as divinities.

In the course of these wanderings they crossed
Lakes Huron and Michigan and arrived in the land
of the Potawatomis in the fall of 1658. These
Indians were a very hospitable and generous tribe,
quite intelligent and affable. They lived on Washington Island and the adjacent parts of what is
now Door County. According to all accounts,
they "loved nothing better than to entertain
strangers who passed that way." Pressed by their
hospitality, Radisson and Grosselliers spent the
entire winter with the Potawatomis, thus becoming the first white men who sojourned for any
length of time within the limits of Door County.
Radisson does not tell much of their experience
here, but notes that it was a succession of feasts
and that the time was spent with "a great deal of
mirth."

"The Potawatomis make their cabins of apaquois (mats), which are made of reeds. All this
work is done by the women. This nation is well
clothed like our savages living at Montreal. The
only occupation of the men is to hunt and to
adorn themselves. They use a great deal of vermillion. They also use many buffalo robes, highly
ornamented, to cover themselves in winter; and
in the summer they adorn themselves with red
and blue cloth. They play a great deal at lacrosse,
twenty or more on each side. Their bat (*crosse*) is
a sort of a small racket, and the ball with which
they play is of a very heavy wood, a little larger
than the balls we use in playing tennis. When
they play they are entirely naked; they have only

a breech-clout, and shoes of deer-skin. Their
bodies are painted all over with all kinds of colors.
There are those who paint their bodies with white
clay, applying it to resemble silver lace on the
seams of a coat; and at a distance one would take
it for silver lace."

They play for large sums, and often the prize amounts to
more than eight hundred livres. . . . . They set up two goals
and begin their play midway between. One party drives the
ball one way, and the other in the opposite direction, and
those who can drive it to the goal are the winners. All this is
very interesting to behold. Often one village plays against
another, the Potawatomis against the Ottawas or the Hurons,
for very considerable prizes. The French frequently take part
in these games. The women work in the fields, raising very
fine Indian corn, beans, peas, squashes and melons. In the
evening the women and the girls dance. They adorn themselves liberally, grease their hair, put on white chemises, and
paint their faces with vermillion, also putting on all the porcelain beads they possess. So after their fashion they look
very well dressed. They dance to the sound of the drum
and rattle which is a sort of gourd with pellets of leads inside.
Four or five young men sing and keep time by beating the
drum and rattle, while the women dance to the rhythm and
do not miss a step. This is a very pretty sight and it lasts
almost all night. Often the old men dance the *medilinne*
(medicine-dance); they look like a band of sorcerers. All this
is done at night. The young men often dance in the daytime,
and strike at the posts. It is in this dance that they recount
their exploits. On such occasions they also dance the scout
dance.[2] They are always well adorned when they do this.

[2] Charlevoix describes these dances in his *Journal Historique*, pp. 296-297. The first of these he calls the "Calumet
dance." Each warrior strikes the post with his hatchet, and
relates his war-like deeds. Of the Scout dance Charlevoix
says: "It is a lifelike representation of all that is done in a

. . . . When this nation goes hunting, which is in autumn, they carry their apaquois with them in order to make their cabins every evening. All the people go, men, women, and children; and they often pass the winter in the woods and return in spring.[3]

NICHOLAS PERROT

It was not until thirty years after Nicolet's visit that the French governor general at Quebec attempted to resume the work of developing the western trade begun by Champlain. In 1665 Nicholas Perrot, an interpreter and agent of the governor, was sent to the great lakes to make treaties with the Indians and open the fur trade again. Perrot was a very able and resourceful man. Though small in stature, he had a commanding personality and a gift of picturesque, forceful oratory which prompted the Indians to regard him with the greatest respect. On learning that the Potawatomis were the most important tribe in the region of the lakes, he went to them and made his headquarters on Washington Island for a year or two.

When he first appeared among them he was received like a supernatural being. They offered him incense, and praised the sun and sky that one of the celestial beings had deigned to visit them.

hostile expedition; and since, as I have already stated, the savages usually aim mainly to take their enemies by surprise, it is doubtless for that reason that they have given to this exercise the name of Scouting."

[3] From a MS. in archives of the Ministere des Colonies, Paris; quoted in *Wis. His. Coll.*, XVI, pp. 366-368.

"Thou art one of the Chief Spirits," they said,
"since thou usest iron. It is for thee to rule and
protect all men. Praised be the sun, which has
instructed thee and sent thee to our country!"
When he wished to walk about they insisted on
carrying him in a blanket. Young men were sent
ahead to clear the path and break away the limbs,
and the women and children fell on their faces to
the ground, not daring to look upon him as he
passed. While refusing this homage as much as
possible, he was able to keep the respect which
inspired it. When he first appeared among them,
most of their young men had started on their
first expedition to Montreal to trade with the
French, of whom they had heard through other
tribes. Shortly after they had started, war broke
out between the Potawatomis and the Menominees, who had their village two miles up the Menominee River. The Potawatomis, left defenceless,
were therefore much afraid of being attacked by
their enemies. But Perrot, hearing of this, went
to the Menominee village and there by his elo-
quence and a few presents succeeded in restoring
peace and good fellowship.

The Potawatomis were very much concerned
about the fate of their people who had gone to
Montreal. They feared that they might have been
overcome by the Iroquois, or that the French had
ill treated them. They therefore asked Perrot's
guide, who was a juggler and medicine man, to en-
lighten them. The juggler built himself a tower of
poles and chanted several songs in which he invoked

all the infernal spirits to tell him where the Potawatomis were. The reply was that they were at
Manistee River, three days' journey away and
were bringing home much merchandise. This
oracle was loudly acclaimed, but Perrot cast a
damper on their spirits by calling the man a liar.
The old men now begged Perrot to tell them when
their people would return. He replied that such
knowledge belonged only to God, but made a
calculation of the length of the journey and their
probable stay in Montreal, and said that the party
could be looked for in about two weeks. There was
now much excitement to see which was the better
fortune teller, the conjurer or Perrot. Fifteen
days later the canoes of the Montreal expedition
were seen in the distance, their occupants firing
off salvos of musketry accompanied by shouts and
yells. When they were several hundred feet from
the shore the two parties began to harangue each
other. Those on shore told with no small exaggera-
tion of the exploits of the great Frenchman who
was among them. Upon hearing this the boat
party, dressed in their French finery, jumped into
the water and swam to the shore to greet him.

Messengers were sent to all the tribes round
about, even to the distant Illinois and Miamis
living hundreds of miles away, to come to the
islands and trade with them and see the great
Frenchman. The next year these Indians came,
thousands of them, and camped at the south end
of Green Bay. The Miamis came with no less than
three thousand men led by a great chief who ruled
over them with much pomp, like a veritable king.

There was also a number of Indians of other tribes
with many great chiefs. Perrot met them as they
arrived, and was treated with the most profound
reverence. Grand feasts, celebrations, and cere-
monies followed. Perrot made a rousing speech
and told them how the soldiers of France would
smooth the path between the Algonquins and
Quebec; would brush the pirate canoes from the
lakes and rivers; would leave the Iroquois no
choice but tranquillity or destruction. To all of
which his Indian auditors shouted "Ho! Ho!"
with the greatest approval. Treaties of friendship
were made, and arrangements to promote the fur
trade. Perrot then returned to the Potawatomis
where he organized an expedition of nine hundred
men, made up in part of other tribes, to go with
him to Montreal with a large cargo of peltries.[4]
Perrot was the most useful man that the French
ever had in the West. He had such commanding
ways about him that the Indians seemed to
respect him more than an army. Whole tribes
plotting mischief fled at the news of his coming.
Several times he unexpectedly entered hostile
Indian villages about to go to war and rebuked
them like dogs. He would lay his breast open
before them and defy them to kill him. His
personal magnetism and resourcefulness so won
the Indians that they sung the Calumet to him
(the greatest honor that Indians could show),
and treated him as the super-lord of all the Indian

[4] From La Potherie's *Amerique Septentrionale*, pp. 85-118,
where many interesting events of Perrot's visit among the
Potawatomis are recorded.

tribes. For forty years, he was the dominant
figure among the Indians of the West. His wonderful bravery, ingenious stratagems and magnetic
oratory saved the French missions and fur trading
stations from extermination on more than one
occasion. In these hostilities, however, his old
friends the Potawatomis on the peninsula were
always faithful to the French.

With Perrot begins an era of great exploration
which made the wealth and vast area of the
western world known to the French. A great
procession of dauntless men were soon pushing
westward, blazing the way for future civilization.
Among these noble men we see the ardent Marquette the discoverer of the Mississippi,[5] the indomitable La Salle who penetrated to its mouth;
the sanguine Hennepin who found his way to the
plains of Minnesota; Duluth, Le Sueur, Tonty,
La Hontan, Charlevoix, Jonathan Carver, Peter
Pond and a score of other intrepid pathfinders.
All these famous men paddled their way past the
palisaded cliffs of Door County. They camped on
its shores, they fished in its waters, they hunted
in the interior for their sustenance, they admired
its bold headlands and its beautiful bays. And,
while lingering here on the threshold of their
enterprise, they reviewed while in the shelter of
Door County's majestic cliffs the details of their
lofty purpose.

[5] The great river had been discovered by the two adven-
turers, Radisson and Grosselliers fifteen years before, but this
fact was not known until a hundred years later.

CHAPTER IV

THE FRENCH MISSIONARIES

Though I walk in the valley of the shadow of death,
Yet shall I have no fear:
For thy rod and thy staff they comfort me,
And thy word shall be my cheer.

DAVID.

One of the inspiring visions of American history is the sublime devotion of the Jesuit missionaries in carrying their gospel of salvation to the savages and cannibals of the western forests. They counted not their lives dear, but forsook kindred, friends, the comforts of home, and the blessings of civilization to devote their lives to the uplift of a savage and degenerate people, a thousand years behind them in manners, morals and intelligence. They followed the Indian to his hunting grounds, threading forests, swimming rivers, camping with them in the somber wilderness, "in weariness and pain, in watchings often, in hunger and thirst, in fastings often, in cold and nakedness." The supposed conversion of a single Indian to the doctrines of the Catholic faith, the baptism of a single infant, seems to have been to them an ample reward for all their labor, and for all their suffering. From the slight memorials which have come down to us, of the labors of love of these venerable and devoted sons of the church, it is evident that no sacrifice was too hazardous, no toil unendurable,

48

which led to the accomplishment of the great
object upon the success of which they had periled
their all in this life. With an ardent devotion to
their duty they thought only of that crown of
glory in the next, which they felt sanguine would
be the reward of their apostolic labors here.
"I have been most amply rewarded for all my
trials and sufferings," says one of the lowly
followers of Jesus, after having, for six days, lived
on a certain kind of rock moss and a part of an
Indian moccasin, given him by a squaw. "I have
this day rescued from the burning an infant who
died from hunger, its mother's resources in the
general famine having failed her. I administered
to the dying infant the sacred rites of baptism;
and, thank God, it is now safe from that dreadful
destiny which befall those who die without the
pale of our most holy church." With us in the
latter days, differing as many do in religious
opinions from this school of ecclesiastics, it is
almost impossible to do them justice.

Judge John Law in commenting on the fortitude
of these devoted men says: "The pioneers in this
great and benevolent enterprise were, like the first
discoverers of fire, morally certain of bringing
wrath on their own heads, and of being condemned
to have their vitals gnawed by the flame of the
funeral pyre, with no eye to pity, no arm to save,
and supported alone by that enthusiasm, self-
devotion and patience under their suffering, which
so eminently characterized these good and holy
men. Death for them had no sting, the grave no

victory. Kissing the symbol of their faith . . . .
they literally gave their 'dust to dust, and ashes
to ashes'; put off mortality to put on immortality;
and with the *Te Deum laudamus* issuing from
their parched lips they laid down their lives in
the wilderness, their requiem the crackling of
fagots, their funeral anthem the war-whoop of
the Indians."[1]

Among these intrepid missionaries of the West
was Father Claude Allouez, one of the most capable and successful, as well as one of the earliest.
Only one missionary had preceded him, the
pathetic figure of Father Rene Menard who, in
1660, made an unsuccessful attempt to establish
a mission on the shore of Lake Superior. He
perished a year afterward in some unknown spot
of the Wisconsin forest, but whether by privation
or perfidy is not known. Father Allouez was his
successor and succeeded in establishing a mission
at the head of Chequamegon Bay near the present
site of Ashland, Wis. A large concourse of people
was gathered there one autumn, among them being
three hundred Potawatomis from Washington
Island and the nearby region. These Potawatomis
begged him to return with them, not so much
because they wished to hear his gospel, but because they desired his help in dealing with certain
grasping fur traders who were among them.

Allouez accepted their invitation, and in 1669

[1] From address delivered before the Young Men's Catholic
Literary Institute, Cincinnati, Jan. 31, 1855, and reprinted
in *Wis. His. Coll.* III, pp. 89-111.

made his journey to the present Door County.
Here he found several villages of Potawatomis on
the islands and in the northern part of the peninsula. He found eight French fur traders among
them who were exerting a very corrupting influence on the Indians. He describes the Potawatomis as being the most docile and generous of all
the Indians he had seen. "Their wives and
daughters are more modest than those of other
nations. They observe among themselves a certain
sort of civility, and also show it toward strangers,
which is rare among barbarians."

Allouez was so hopeful of his work among the
Indians of the peninsula that he established a
mission among them. This was the mission of
St. Francis Xavier. On his excellent map, made
in 1670, he shows the mission to have been on the
east shore of Green Bay, near the south-east corner
of present Door County. One year later this
mission was removed to the present site of DePere,
six miles from the head of Green Bay, where a
substantial chapel was built. Father Louis Andre
now arrived and became his regular assistant. In
the next two years they baptized more than two
thousand Indians.

Allouez and Andre now divided the work,
Allouez extending his missions further south, while
Andre devoted his labors to the Indians near the
mission and northward on the peninsula. He was
very successful, and soon gained five hundred
church members. In lieu of chapels, wooden
crosses were erected and dedicated in a number of

places where mass was said and the Indians gathered for prayer. One of these crosses was erected on the neck of land between Ellison Bay and Rowley's Bay, and was maintained for almost two hundred years. When D. H. Rice, the first settler, took possession of this land in 1857 he sometimes saw Indians gather there and make their invocations.

The work begun by Father Allouez was so successful that it was necessary for many other missionaries to join him. Among these were Fathers Silvy, Albanel, Chardon, Dablon and others. Indians by the thousand were baptized and the missionaries for a time were treated like divine beings. So also was the chapel. The Indians sometimes gathered around it by the hundreds invoking it as they did their idols and seeking to propitiate it by gifts of tobacco with which it was flanked on every side.

In the last days of October, 1676, Allouez set out for a visit to the Illinois Indians. He left the mission in a canoe with two companions intending to paddle his way along the shores of Green Bay and Lake Michigan. The winter came unusually early that year and they "had not gone far before, the ice overtaking us, we were compelled to go into camp and wait until the ice was strong enough to bear us. It was not until the month of February that we began (resumed) our voyage, a very unusual mode of navigation, for, . . . we placed it upon the ice, over which the wind, which was in our favor, and a sail made it go as on water."[2]

About thirty miles north of the mouth of Sturgeon Bay, he passed near a village of Potawatomis, no doubt at Ellison Bay, and learned that a young Indian whom he had previously baptized had been killed by a bear. He stopped to make prayers for the dead and was then persuaded to remain for a time. During his stay a great bear hunt took place among the Indians, which he describes:

Afterward by way of avenging the death (of the young Indian), the relatives and friends of the deceased went to make war on the bears while they were still in good condition, that is to say, in winter; for in the summer they are so thin and so famished that they even eat toads and snakes. The hunt was so successful that, in a short time, they killed over five hundred, of which they gave us a share, telling us God delivered the bears into their hands as satisfaction for the

[2] Anyone familiar with ice conditions in Green Bay will recognize the circumstances which obliged him to stop. The first ice-formation of the winter consists of rough slush ice which packs into the deeper coves and harbors, while the main body of the bay remains free from ice for a long time afterward. Allouez and his companions must have camped one night in such a harbor. In the morning they were unable to reach the open water outside, because of the roughness and insecurity of the slush ice. The only harbors within the limitations of his course and the extent of his journey, and deep enough to have hindered them, are those of Egg Harbor and Fish Creek, and in one of these places he was obliged to camp for three months. It could not have been in the southern part of Green Bay, because here the ice forms early. Moreover, if hindered at this point, he would no doubt have returned to the comforts of the Mission station.

death of that young man who had been so cruelly treated by one of their nation [the bears].[3]

From these Indians Allouez probably learned that there was a safer and a shorter way to reach Lake Michigan, southward bound, than by the dangerous strait at Death's Door; for we see him retrace his course and land at Sturgeon Bay, from which there was a portage four miles ("a league and a half") long to Lake Michigan.

[3] *Jesuit Relations*, LX, page 151 and *ante.* Pierre Le Sueur, a prominent fur trader, describes the hunting of bears as follows:

"The bears climb up hollow trees, and wedge their bodies into the hollow places of these trees. They remain there six or seven months without leaving their refuge, and nourish themselves only by licking their paws. When they enter those holes, they are very poor in flesh; but when they leave them, after winter has ended, they are so fat that they have a layer of tallow half a foot thick. It is almost always in the poplar or cypress that the bear hides, because those trees are usually hollow. When men undertake to kill them, they place against the tree in which the bear is, another tree, which extends to the hole by which the bear entered. A man ascends by this latter tree, and through the hole flings into the hollow some pieces of burning wood, which compels the animal to come out, in order to escape being burned. When it has emerged from the hollow of the tree, it descends backward, as a man would; and while it descends the hunters fire their guns [or arrows] at the beast. This is very dangerous hunting, for although the animal may be wounded, sometimes with three or four gunshots, it will still hasten to attack the first person whom it encounters, and in an instant rends them, with a single blow of teeth and claws. There are bears as large as a coach-horse, and so strong that they can easily break a tree as large as a man's thigh." Margry's *Decouv. et etabl.* V, pp. 419-420.

For about ten years the missionaries met with much success. Then a reaction set in. The novelty had worn off, the miracles did not seem to happen so often, and the inherited respect for Indian mysteries began to reassert itself. But the principal obstacle to missionary success was the debasing influence of the fur traders. These men carried brandy with them as the most effective means to make a profitable bargain with the Indians. Debased by drunkenness the Indians became insolent and brutish. In vain the pious missionaries exhorted their proselytes to lead a virtuous life. The degrading influence of scores of unscrupulous fur traders nullified all their efforts. We see the despair of the anxious fathers reflected in the scathing denunciation of the fur traders, written by the venerable Father Chardon, to the French governor.

These traders, he writes, "have introduced the two infamous sorts of commerce which have brought the missions to the brink of destruction . . . the commerce in brandy and the commerce of the savage women with the French. Both are carried on in equaly public manner, without our being able to remedy the evil because we are not supported by the commandants. . . . All the villages of our savages are now only taverns, as regards drunkenness, and Sodoms, as regards immorality, from which we must withdraw, and which we must abandon to the just anger and vengeance of God."

About 1680 in a drunken brawl an attack was made on the mission at DePere and several members of the mission were killed. When the Indians became sober they were greatly frightened, fearing swift vengeance from the great father in Quebec. The missionaries, however, exemplifying, the doctrine of forgiveness, demanded no punishment, and no vengeance came. The Indians misunderstood this tolerance, despised the French for weaklings, and matters grew worse. The Jesuits were subjected to all sorts of indignities. Once an Outagamie chief brought all his old and dull axes to the mission and compelled the reverend father to temper and sharpen them, meanwhile standing over him with a tomahawk. When the work was done, instead of showing gratitude, the chief gave the father such a beating that he was obliged to keep to his bed for many days.

About this time a plague broke out among the Indians around the bay which greatly diminished their numbers. The Indian sorcerers, after many incantations, claimed that this plague was brought on by the witchcraft of the "black gowns" (Jesuits), in revenge for their injuries, and urged that they be exterminated with all the French. For a while it looked as if a general massacre would follow. Only the adroit diplomacy of the redoubtable Perrot prevented it. But when he retired from active service a few years later his great, restraining influence by which the Indians had been held in check for forty years, was broken. Intertribal wars broke out and continued until one side or another was practically exterminated.

Meanwhile the missions were destroyed and abandoned, and the fur trade languished. The Indians turned again to their idolatrous practices and their last stage was worse than their first.

CHAPTER V

THE FIRST PIONEER

Not for us delectations sweet,
Not the cushion and the slipper, not the peaceful
and the studious,
Not the riches safe and palling, not for us the tame
enjoyment,
Pioneers, O pioneers.

WHITMAN.

Once upon a time there was a man by the name of Increase Claflin. He was the first white settler on the Door County peninsula.

Claflin was a worthy representative of that indomitable advance guard of American pioneers who, compelled by mixed motives, fearlessly penetrated into the unknown wilderness to blaze a trail for the civilization that followed.

He was born September 19, 1795, at Windham, N. Y., and was a descendant of a long line of sturdy Yankees who, centuries ago, conquered the forests of New England. His father, also named Increase Claflin, was a member of the Hopkinton Company of Minutemen who responded to the Lexington alarm. He served with honor throughout the Revolutionary War.

His grandfather, Cornelius Claflin, was a soldier who had served in the French and Indian War. Later he also served throughout the Revolutionary War as lieutenant in the same regiment in which his two sons were privates.

58

Like his father and grandfather, our Increase Claflin was also a soldier. In 1812, when less than seventeen years of age, he enlisted and served through the war with Great Britain. But, unlike his ancestors, Increase Claflin was of an adventurous, roving disposition, and about 1820 we find him spending a few years among the fragrant magnolias of New Orleans. Ten years later he is chasing the redskins of southeastern Wisconsin in the so-called Black Hawk War.

Some time before this he was established as a fur trader at Kaukauna, Wis. Here he must have had a number of men in his employ, for the census of 1830 gives the number of people in his household as thirteen.

That large, nameless peninsula to the northeast of Green Bay, with its bold promontories and mysterious coves, which the Indians described as their original paradise, soon attracted him. After visiting it he decided that here was the most attractive region he had seen in his many thousand miles of wanderings. On March 19, 1835, he set out to blaze a new trail for civilization to follow. A trackless forest jungle of forty miles lay between him and his destination, but Claflin found an easier way. Old Jack Frost had prepared a smooth highway for him on the ice of the bay. On one sleigh was a load of hay. On another was his sailboat in which was stowed his family, his furniture, tools, grain, provisions and other necessaries, while behind followed his cattle and some breeding horses. Thus equipped he made his way to Little

Sturgeon Bay. On the point of land at the mouth of this bay, on the west side, he built the first white man's house on the peninsula.

Here Claflin lived an independent, masterful life. He produced on his farm practically everything that his family ate and wore. Emergencies, such as sickness, fire and protection of his homestead from prowlers, he met for himself. Besides this he exported large quantities of food supplies and other things needed in the world he had left behind. He paid no taxes and he needed none to help him provide peace, order and progress in his own field of activity.

Little Sturgeon Bay was then as even now a most idyllic spot, abounding in all kinds of fish and game. Along the west side of the bay stretched miles of grassy marshes and meadows, affording good forage for horse raising, which was Claflin's principal purpose. On the opposite side of Little Sturgeon Bay, on what is now called Squaw Point, was a village of four or five hundred Menominees. Claflin got along very well with these Indians, as he treated them fairly and generously. Two or three years later, however, serious trouble broke out between them, brought on by Claflin's son-in-law, Robert Stephenson.

This man, originally from Pennsylvania, came to Little Sturgeon in 1836, and was employed by Claflin in various capacities. The next year he married Claflin's oldest daughter, but continued for a time to make his home with the Claflins. He was a capable, energetic man, but domineering and tricky in dealing with the Indians. His usual procedure was to get the Indians drunk, whereupon he would obtain their peltries at prices ruinous to the poor redskins. This greatly displeased Claflin, who was as fair to the Indians when drunk as when sober.

One day, when Claflin returned from a round-up of his horses, an alarming sight met his eyes. A band of Indians in war paint were scurrying around his cabin. Stephenson was engaged in a hand to hand fight with several redskins and was felled to the ground with several knife stabs. Another white man in the employ of Claflin lay dead in the doorway, and a couple of Indians were just dragging out his daughter, Mrs. Stephenson. Dashing his horse into their midst, Claflin scattered the Indians who were dragging away his daughter, and hurried her into the house where he found the other members of his family safe, but trembling with fear. Turning to the Indians he demanded the meaning of the attack.

A stalwart Indian, their chief, stepped up and spoke:

"You are our friend and we wish you no harm. You may take your squaw and your papooses and go away in your boat. But we shall kill your son [pointing to Stephenson], and burn your house and let no white man stay here among us. Our young men bring their furs and our daughters their robes and blankets to your house and he [Stephenson], makes them drunk with firewater and gives them nothing in return. We shall kill him and give his squaw to our young men for our daughters to laugh at and spit upon. Go, while we remember your good deeds!"

In vain Claflin tried to reason with them, but a hubbub of excited Indian outcries and threats broke out. Claflin then said:

"Well, if I have to go, let me treat you before I go. We have always been friends and let us part in the same manner."

The Indians grunted their approval of this and seated themselves in anticipation of their feast.

Claflin then entered his storehouse, and returned with a keg and a tin cup. He carried it into their midst and poured a little of the liquid into the cup. To their amazement the Indians saw, not whiskey, but gunpowder trickle into the cup. Then he took his flint and fire-steel, ignited a piece of tinder and threw it into the cup. There was a flash, a loud report, and the cup was gone!

Apprehensively the Indians looked at each other and fidgeted in their seats. The chief then said, "Wh-what is my white brother going to do with the keg of powder?"

"Do!" exclaimed Claflin, "I am going to blow you all to hell! Either you smoke the peace-pipe with me or not a man leaves this spot! I have aways treated you Indians fairly and squarely and now you turn upon me like wolves to kill my children and drive me from my home. If my son Robert has misused you, you have punished him enough. Now let us be friends and smoke the pipe of peace."

Filled with mixed feelings of admiration and apprehension at Claflin's audacity, the Indians assented. Claflin filled his pipe and lit it, where-

CLAFLIN'S ELMS AT LITTLE STURGEON BAY

upon it was passed from Indian to Indian with all proper solemnity. Two gigantic elms near the shore at the mouth of Little Sturgeon Bay now mark the spot where this eventful meeting took place.

The Indians made no further trouble for Claflin and his household, but the strained relations between Claflin and Stephenson increased. Finally, like Abraham of old, Claflin decided to leave his son-in-law in possession of the favored land, and with his family go elsewhere to seek a home. In 1844 he went twenty miles north and settled on a promontory one-half mile north of the present site of Fish Creek. This promontory, now embraced in Peninsula Park, is still known as Claflin's Point, and is the most popular camping ground in the park.

Increase Claflin was a splendid type of a pioneer. He was reliable, fearless, resolute, loyal and self-sacrificing. In the rare quality of his ancestors as well as in his own sturdy manhood, Door County could ask for no truer type of American virtue. There is a familiar painting of fine conception typifying "The Spirit of 'Seventy-Six." Three figures of martial bearing are seen advancing at the head of a body of troops. In the middle is the grandfather, white locks flowing in the wind, playing on a flute. On one side is his son, a drummer in the prime of life. On the other side is the grandson, not yet fully grown, but catching inspiration from his elders and keenly beating his drum. Advancing onward they make a soul-stirring picture.[1]

In the history of the Claflin family there are events that are just as soul-stirring as this famous

[1] The painting referred to is by Archibald M. Willard. The picture is now in Abbott Hall, Marblehead, Mass.

painting. As a parallel we see Increase Claflin's grandfather, the Revolutionary lieutenant, charging the breastworks of Crown Point, closely followed by his two sons. By such was America freed! And as a climax we see Increase Claflin, the Door County pioneer, now old and weary of days, standing in the doorway of his Fish Creek cabin, speeding his three sons to war for the preservation of his country. In the summer of 1862, when the President called for troops to save the Union, Claflin sent his three sons as volunteers, saying: "If I had twenty more, they should all go!"

CHAPTER VI

A FORGOTTEN COMMUNITY: A RECORD OF ROCK ISLAND, THE THRESHOLD OF WISCONSIN

There is a pleasure in the pathless woods,
There is a rapture on the sounding shore;
There is society where none intrudes
By the deep sea, and music in its roar.

BYRON.

Off the exteme northeastern corner of Wisconsin lies a little island about a mile square. It is situated in the middle of the mouth of Green Bay, storm-lashed by all the heaving seas of Lake Michigan. On the north and west its castellated limestone ramparts rise in perpendicular grandeur from the lake to the height of a hundred feet and more. On the south and east, however, its shores slope gently down until their sands blend with the lapping waves of the inland sea. From shore to shore the interior is now covered with a majestic mantle of forest green, shrouding a solitude which for fifty years has been unbroken by human habitation. Only upon the northern cliff sits a watchful lighthouse keeper, turning his gleaming light throughout the night upon the dark waters to warn away the wind-swept mariner from the dangerous coast he is guarding.

Seventy years ago this isolated little Island, now ruled over only by the "murmuring pines and hemlocks," was the home of an energetic community of about a hundred people. Their snug homes lined the eastern shore and their sailboats ventured far out to sea for fish and fun. Up on the hillside a number of early Wisconsin pioneers are laid away to rest, and in a log schoolhouse whose very site is forgotten many worthy citizens of this State and Michigan have learned their A B C's.

Rock Island, the subject of our sketch, was well known to the early French explorers under the names of Potawatomi Island or Louse Island.[1] It is without doubt the first place in Wisconsin visited by white men. When Jean Nicolet in 1634 passed through the Straits of Mackinac, the customary Indian route was along the shore of the northern peninsula of Michigan until the present Point Detour was reached. There the natives crossed the mouth of Green Bay to the north shore of Rock Island and Washington Island and then followed the west shore of the Door County peninsula to the Winnebago capital at Red Banks.

The first permanent American settlers on Rock Island were John A. Boone, James McNeil, George Lovejoy, David E. Corbin, Jack Arnold, and Louis Lebue. Most of these were fishermen and trappers who came from the island of St.

[1] The name Potawatomi Islands was often applied to the entire group including the present Washington Island. The term "Louse" is a corruption of the original French name. The French abbreviated the word Potawatomi (often spelled by them Poutouatami) to *Les Poux*, by which they meant the Indian tribe, not the insect.

Helena in the Straits of Mackinac in 1835 or 1836.
As they were the first settlers in the northeastern
part of the State outside of the settlement at
Green Bay, a brief mention to their personalities
will be desirable.

John A. Boone was a quiet, apt-spoken man
who, without thrusting himself forward, was
always looked upon as the leader of the community

ROCK ISLAND

that grew up on the Island. He had evidently
spent his entire life on the frontier, as he spoke the
Chippewa dialect like a native and fully understood the Indian character. These accomplishments later served him well when he was the means
of averting a very threatening Indian war. He
was a married man when he settled on the Island,
and lived there until his death in 1866, when he
was fifty-two years old. A little white-painted
cedar cross still marks his grave on the Island.

George Lovejoy had been a sergeant in the
United States army, having seen five years' service

at the frontier post of Fort Howard, during which
he had taken part in expeditions of various kinds
to the Indians. He was a hunter of fame in many
parts of northeastern Wisconsin and an eccentric
bachelor of remarkable capacity for almost anything he undertook. He could beat an Indian on a
trail, and he astonished the sailors by building on
Lake Michigan one of the best schooners with
which he traded along the shore. His commercial
qualities were crude, however, and barren of
success. He was an expert with the violin and a
master ventriloquist. Sometimes he would go out
on the ice when an Indian was fishing and make
the trout talk back to its captor in the most
approved Chippewa dialect, to the poor Indian's
terrorized amazement. This, with his reckless
bravery and easy skill in every undertaking, made
the Indians look upon Lovejoy as a veritable
demon, and they were always most anxious to
propitiate his favor by gifts of all kinds. In one
direction, however, Lovejoy was anything but
brave. That was in his attitude toward the fair sex.
When suddenly confronted by a woman he was
struck dumb with embarrassment and often fled
precipitately. This failing of his was the cause of
many broad jokes played on him by the mischievous young folks of the little community.

To James McNeil belongs the honor of being the
first taxpayer in Door County. He owned the
entire south shore of Rock Island. He was an old
bachelor of a penurious disposition, with a failing
for whisky. He was very close-mouthed about his

own affairs except when the jug arrived from
Chicago. Under its stimulus he would become
confidential and would prate with tipsy garrulity
of his "yellow boys," which, he confided, would
support him in comfort when he should retire.
By "yellow boys" he referred to his store of gold
coin, which, unfortunately, became his undoing
instead of his support. One morning the poor old
man was found beside his chicken coop wounded
and unconscious. When he came to, he shouted,
"Boone! Boone!" in agonized appeal. Boone, who
was justice of the peace, was quickly summoned,
but by the time he appeared McNeil had passed
away, taking the secret of his murder with him.
No positive clue to the murderer was ever obtained, but it was believed that a strange craft that
had been seen in the vicinity contained the crimi-
nals. For some time there was much hunting in
the potato patch and among the crags for the old
man's treasure but nothing was found.

Both David E. Corbin and Jack Arnold were old
soldiers who had been sergeants in the War of 1812.
Corbin was the first lighthouse keeper in Wisconsin, being in charge of the Rock Island lighthouse
(the first in Wisconsin) from its construction in
1836 until his death in 1852. Arnold stayed with
Corbin in the lighthouse because they were such
inseparable cronies. They rarely ever conversed
but were apparently able to read each other's
thoughts. When finally Arnold sickened and died
in 1848 Corbin watched by his bedside with ceaseless vigilance, caring for him with the greatest
tenderness.

Of all these men Louis Lebue is the only one
from this section whose name is mentioned in the
territorial census of 1836. In 1843 he had the
misfortune to lose his wife, who was buried on the
Island. This unsettled him, and he departed for

OLDEST HOUSE ON PENINSULA, BUILT ON ROCK
ISLAND IN 1836

Chicago, the rising metropolis of the West. On
Calumet River, near Chicago, he made the
acquaintance of some men by the names of Miner
and Luther. Henry D. Miner was the son of a
clergyman who, as early as 1828, had settled at
Kaukauna as a missionary among the Indians.
The following year he died of fever at this place.
His boy, Henry, who was then eight years old,
returned to his relatives in New York. In 1842,

however, he returned to the West accompanied by his brother, T. T. Miner, and Job, Seth, and Brazil Luther. In the spring of 1844, Lebue met these men and told them of the easy living that could be made on Rock Island by fishing. He showed them how to repair and knit twine, and initiated them into the mysteries of the piscatorial art. As a result he sold them his outfit, whereupon in June, 1844, they moved up to Rock Island to become the forerunners of a steady advance of settlers to this distant region. Job Luther had a vessel and at intervals he freighted fish down the lake and fishermen up, until after three or four years there were upwards of fifty men, many of them having families, living on Rock Island. Nearly all of these people came from Lemont, near Chicago, and were known as the Illinois Colony. Among them was a number of sturdy pioneers by the names of Chauncey Haskell, Robert, Sam, and Oliver Perry Graham, and, last but not least, old Father Kennison.

Old David Kennison was for a while the most famous character on Rock Island, entertaining its denizens with tales of stirring events in the infancy of the Republic in which he had personally taken part, for he was more than a hundred years old. He could say with Tennyson's Odysseus:

Much have I seen and known; cities of men, and manners, climates, councils, governments, and drank delight of battle with my peers—

Kennison was the last surviving member of the Boston Tea Party. He had participated in several wars and smelled blood in many battles. Now, after a century of toil and trouble he had come to Rock Island, satisfied to ruminate in peace upon a busy life.

He was born in 1736 in one of the frontier settlements of New Hampshire. Of his youth and early manhood we know nothing. Very likely he carried a musket in the French and Indian War and had his share of fighting against the Indians in the region of the home of his youth.

At the outbreak of the Revolutionary War we find Kennison right in the midst of things. He was a member of the Boston Tea Party, a participant in Lexington and Bunker Hill and many other battles of the Revolution, surviving them all. Keenly loving the strenuous life, he later went west and in 1804 enlisted at Chicago, serving for eight years at Fort Dearborn until the massacre of the post drove the surviving numbers of the garrison east.

Having a charmed life and not yet sated with fighting, he then served through the War of 1812, escaping as before without a scratch.

Being eighty years of age Kennison now settled down to a quiet life, but here his expectations miscarried. He should have remained in the army, for he discovered that a civilian's life was vastly more dangerous. "A falling tree fractured his skull and broke his collar bone and two ribs; the discharge of a cannon at a military review broke both of his legs; the kick of a horse on his forehead left a scar which disfigured him for life." Trouble thickened fast and thickened faster.

In spite of all his public and private tribulations
Kennison succeeded in marrying four times and
becoming the father of twenty-two children. The
records are silent about these successive broods
for they faded away, until Kennison in his old
age had only one son to lean upon. With him at the
age of one hundred and ten years he went to Rock
Island to repair twine and clean fish.

Kennison found life on Rock Island very
pleasant. The big sea surging all around him from
beyond the skyline spoke to him of other days and
events and was soothingly companionable. Up at the
lighthouse were David Corbin and Jack Arnold,
two other veterans of the War of 1812, with whom
he now and then swapped reminiscences. The old
man was almost happy.

But not for long. After some years the twenty-
second son got tired of the island and joined the
other twenty-one renegades in deserting their
father. When the winter came, the fishermen
scattered, as was their wont, to different cities to
spend a part or all of their earning in having "a
good time." Old man Kennison was obliged to
leave and secured free passage to Chicago, hoping
to eke out a meager existence on the paltry pension
of eight dollars a month which the government
doled out to this last surviving champion of her
liberties.

As this pension was insufficient he was finally
constrained to enter a public museum and obtain
a small pittance by exhibiting himself as a curiosity. Finally he died February 24, 1852, one

hundred and sixteen years of age. At last his troubles were over.

Then, at last, when he was beyond the reach of knowing it, came recognition and honor. On the day before his death, in response to a request presented in his behalf that he be saved from the potter's field, the city council with patriotic promptness voted that a lot and a suitable monument be provided for him in the city cemetery. A grand funeral was held from the Clark Street Methodist Church and many clergymen assisted in the services. The procession moved in two divisions from the church to the cemetery with cannon booming at one-minute intervals. In the procession were the mayor and city council, a detachment of the United States army, various military companies and bands of the city, companies of firemen, and a large part of the population of the city. The cemetery at that time occupied a portion of the ground which later was included in Lincoln Park. When this area was added to the park the bodies interred in it were removed, but Kennison's was left undisturbed. in 1905 several patriotic societies joined in marking his grave by placing upon it a large granite boulder to which is fixed a bronze tablet containing an appropriate inscription.

Here old David Kennison will probably forever lie, in unique dignity, in Chicago's most beautiful park. He was greatly honored in his death, but his declining years were suffered to be spent in cleaning fish and later in being gaped at by the idle and curious multitude.[2]

Nearly all these people lived along the sandy east shore of Rock Island where they constituted the first community on the Peninsula. And a very contented community it was. The fish were plentiful and very large, often only ten to fifteen being required to fill a half-barrel. In the woods was an abundance of game and in the little garden patches of the settlers potatoes and other vegetables grew luxuriantly. Apples, plums and berries in abundance grew wild in the woods, and there was no lack of firewood with which to keep warm in winter time. It was a free and easy life to lead, somewhat indolent and uncouth, without taxes or sociological troubles of any kind. Their chief handicap was their distance from any post office through which to learn the news of the outside world. The most accessible one was Chicago, three hundred miles away. Mail intended for the settlement was usually directed as follows: "H. D. Miner, Rock Island, care of Williams, Chicago, Illinois." On his occasional visits to the metropolis, Job Luther would get the little bundle of Rock Island letters and newspapers, often many months old. On such visits he would also lay in ample stores of tea and tobacco, boots and biscuits, soap, sugar and soda, coffee and calico, and all the other

[2] For further information on David Kennison, see Quaife's *Chicago and the Old Northwest*, pp. 255-257. See also *Chicago Democrat*, Nov. 6 and 8, 1848 and Feb. 25-27, 1852; *Chicago Daily News*, Dec. 19, 1903.

staples which T. T. Miner carried for sale in his little store on the Island. Besides these things he was also entrusted with a multitude of private requisitions, such as a mouth organ or a fowling piece for a young hopeful, or a bonnet or a brocade for one of the fairer sex. Such fineries were needed to do honor to the occasional weddings, funerals, and other events of importance. Weddings were of rare occurrence and while of transcendent interest were usually not attended with any ceremonial, being in the absence of church and organized state only "common law marriages." Now and then a contracting couple was found who felt the need of the blessing of the church upon their union. This, however, was difficult of attainment. On one such occasion H. D. Miner was drafted into service to tie the knot. The cause of his selection was that a certain faint glow of sacerdotal dignity was attributed to him by reason of the fact that his father had died as a missionary to the Indians. Miner complied, and with all the unction he was capable of, joined together Henry Gardner and Elizabeth Roe, the first marriage ceremony to be performed in Door County.

Another wedding is recalled by the old pioneers with much relish. It was a big affair in which two Norwegian couples were joined in wedlock, and fishermen from many shores had gathered to celebrate the double feast of love and liquor. As usual, there was no clergyman to officiate, but a humble visiting evangelist was drafted into service. He had no license to perform a marriage

ceremony, but he was anxious to please his pro-
spective converts and consented to officiate. It
was a new undertaking for him, and being nervous
and not knowing the contracting parties, he made
the unfortunate blunder of marrying the two men
to each other and then the two women. The two
Norwegian bridegrooms on their part had but
little knowledge of the English language and only
a very dim notion of the procedure at an American
wedding. They, however, had a vivid impression
that it was their part to answer "yes" when spoken
to. When, therefore, Ole Olson was asked if he
would take John Johnson for his wife and vice
versa, an energetic "yes" was the response to the
uproarious acclaim of the assembled guests. It
was not until the young exhorter was similarly
joining together the two brides, who, by the way,
were sisters to begin with, that the officiating
witnesses rallied their wits and interposed, whereupon a fresh start was made.

Dependent on the lake as these people were and
exposed to all its squalls, hairbreadth escapes on
the water were quite frequent. While thrilling
adventures were common, the fishermen were so
used to Neptune's whims that comparatively few
fatalities occurred. Now and then, however, one
would be caught unawares and go down to his
watery grave. A notable instance of this was the
drowning of the Curtis family.

Newman Curtis joined the Illinois Colony in the
later forties. In the summer of 1853, he went with
his family, consisting of his wife, daughter, and

newborn baby, to St. Martin's Island to fish. After a successful season he prepared to return in the fall to his permanent home on Washington Island. He was accompanied by his nephew, W. W. Shipman, and Volney I. Garrett, two young boys.

As Mr. Curtis had a quantity of household goods and freight he rented an old heavy-built schooner, which in early days had outridden many a storm but was now considered too unwieldy to be safe. But as it is but eight miles between the two islands the little party started off without fear. All went well until the vessel was drawing quite near to Washington Island where its occupants could almost see their little white cottage among the trees on shore. By this time the fair wind that had favored them had gained in force until a storm was blowing and the creaking old schooner began to roll heavily. In doing this she took in a great deal of water as the top seams were quite open. The pump was kept going but in spite of this the vessel settled fast and soon was so water-logged as to be quite unmanageable. When just outside of Indian Point, on which the seas were rolling terrifically, those on board realized that in all probability the schooner would sink before she would be dashed on the rocks, not a hopeful alternative. Curtis and Garrett, therefore, prepared to lower the yawl while Shipman went down to fetch the baby who was still sleeping in an upper bunk oblivious to its peril.

At this juncture a heavy sea dashed over the

vessel from stem to stern, tearing away the frail
grip of the Curtis girl on the cabin to which
she was clinging, and washing her overboard.
This wave was followed by another which tore
loose the yawl, throwing it into the sea endwise
and pinning Curtis underneath it. When he
finally came to the surface he was so overcome by
his exertions and bruised by the blows he had
received that he was unable to swim the few feet
that separated him from the yawl which floated
away filled with water. Upon seeing sudden
death thus overtake her daughter and husband,
Mrs. Curtis for a moment forgot her own peril
and stretched out her arms to them screaming in
anguish. Instantly she, too, was washed overboard.

By this time Shipman, drenched with water,
had emerged from the cabin with the baby in his
arms. He made for the remaining hatch, reaching
it simultaneously with Garrett, who also seized it.
"Who takes the hatch takes the baby," shouted
Shipman, thrusting the baby toward his companion. Garrett, however, with an oath refused this
handicap. The next moment they were all thrown
into the water. Clinging to the hatch, Shipman
and his charge made land safely, where they were
soon joined by Garrett, clinging to the submerged
yawl. The next morning the battered bodies of
the Curtises were found on the beach and were
buried on Rock Island.

Besides the Illinois Colony and other white
settlers, there were about fifty wigwams of Chip-

pewa Indians on Rock Island, living under the leadership of their renowned chief, Silver Band. The two communities got along very well together except on one occasion when open war was threatened. It happened in this way. Among the whites was a widow by the name of Oliver. She had three boys, one of whom, Andrew, was a half-grown fellow. Widow Oliver was much broken down over the loss of her husband; but was nevertheless in great demand for nursing the sick, at which she was very capable. Her boy one day took her place in the kitchen where he was peeling cold boiled potatoes. Some of the Indian urchins noticed this through the partly opened window, and soon there was a group collected, their noses pressed flat against the glass, making grimaces at the white youth and calling him "squawman." This was too much for the willing Andrew, who suddenly threw a cold potato at the leader of the band of mockers. He, however, dodged the missile which, with splinters of glass, struck an innocent little bystander full in the eye—the seven-year-old son of Chief Silver Band. The screaming sufferer, bleeding profusely, was hurried to his father's tepee, and soon the Indians were seen rushing excitedly back and forth. The white settlers, on hearing what had happened, felt that a crisis was imminent, and sent Henry Miner to parley with the chief. He was met at the door and gruffly told to go away. Others attempted to interview the Indians, but without gaining a hearing. The whites were fast becoming terror-stricken for

they knew that at any moment a signal could be
sent to the neighboring Indians on Washington
Island, and they would have no chance against the
overwhelming numbers that might be brought
against them. Some of the more reckless favored
taking time by the forelock and making a sudden
onslaught upon Silver Band and his people. "If
not," they declared, "we will surely be massacred
in our beds." Others, more timid, recommended
rather that Andrew Oliver be killed and brought
before the enraged chief as a fitting sacrifice. In
the midst of this hubbub John Boone arrived. He
could talk Chippewa fluently and was highly
esteemed by Silver Band. Taking the weeping
Widow Oliver by the hand he made his way to
Silver Band. In well-chosen words he reminded
the chief of their earlier associations. He called up
one picture after another of the chief's greatness
in war, and cunning in battle, and mighty prowess
in hunting the bear and the buffalo. He told of
how wisely Silver Band had conducted the affairs
of his people as chief, keeping them out of trouble
of all kinds, showing magnanimity to his foes,
and gaining the esteem and confidence of the white
people. He concluded:

And now I am glad that so magnanimous a chief as Silver
Band rules his people. Children play, children quarrel,
children get hurt. It is easy to be magnanimous when
another's child is hurt, but not so easy when your own child,
the pride of your eye, suffers. Another chief, less noble than
Silver Band, would let rage master him, and thus bring
everlasting trouble upon himself, his people, and his neighbors.
Not so with my brother, the great chief Silver Band, the lord
of the Chippewas. He suffers, but he forgives.

And now I bring you this woman to be your handmaiden. She is weak of body and crushed with grief that her son should unwittingly have brought this evil upon his little playmate, your son. But her hands are skilled in the mixing of potent medicinal herbs, and she can nurse your child to life.

Soothed, complimented, and exalted by this skillful discourse, the chief sat silent. Finally he rose, extended his hand to Boone, and led Widow Oliver to the couch of his suffering boy. There she remained nursing him unremittingly until he was able to go about again, blind, however, in one eye.[3]

Little by little the fortunes of Rock Island declined. In the fifties and early sixties when other parts of Door County began to be occupied, the exodus from Rock Island began. The Island's lack of good harbors, and the inconveniences attendant upon its isolation, more than out-weighed the greater profits derived from its fishing. One by one the old-timers slipped away to seek their fortunes in other parts. Some of the build-ings were removed while others mouldered away. It is now long since the Island's last loyal denizen bade good-by to his romantic habitation. Where once stood the village of the Illinois Colony wild roses now grow and the rabbits and chipmunks frisk undisturbed over the knoll that marks the site of the old schoolhouse. Up on the hillside lie

[3] When last heard from Andrew Oliver was at the head of a manufacturing establishment in Allegan, Michigan. The Indian boy, Kezias, is now the chief of the same band of Chippewa with headquarters on the peninsula of northern Michigan, between Big and Little Bay de Noquet, where he is also their priest and teacher.

the bones of John Boone, Silver Band, Newman
Curtis, and all the other worthy men who played
a man's part in their day; the moss of the forest
has garbed their graves, and their aspirations and
their deeds are alike forgotten.[4]

[4] Rock Island is now the property of Mr. C. H. Thordar-
son of Chicago, whose solicitious aim is to cherish it and
protect in such a way as its beauty and history deserve.

CHAPTER VII

WASHINGTON ISLAND

Three fishers went sailing out into the West,
Out into the West as the sun went down;
Each thought on the woman who loved him the best;
And the children stood watching them out of the town;
For men must work, and women must weep,
And there's little to earn, and many to keep,
Though the harbor bar be moaning.
                                        CHARLES KINGSLEY.

Far out amid the white-crested waves of Lake Michigan lies Washington Island. Its nearest point is about twenty five miles from the mainland of northern Michigan, while the tip end of Door County peninsula comes within six miles on the south. North, east, and south lie a number of islands, constituting the "islands of Green Bay," and known as a dangerous zone of navigation ever since the first sailing vessel that plowed the waters of the great lakes, the *Griffin* of the famous explorer La Salle, was wrecked there in the year 1679.

Washington Island is about six miles square and has a shore line of twenty-six miles. On the north and west sides, the shores are high and precipitous, particularly at the northwestern extremity. Here Bowyer's Bluff raises its limestone ledge imposingly to the height of almost two hundred feet from the water's edge. These cliffs are seamed with caves and fissures, and carved into fantastic figures by the storms of bygone ages; but now the clinging cedars are weaving a drapery of green for their rugged sides. The south and east sides, on the contrary, are mostly low and sandy, with a shallow water front. On the north side is Washington Harbor, a bay which extends into the island about a mile and a half, and surrounded by sloping, timbered hills. On the south, too, is a large indentation, known as Detroit Harbor, which is made a landlocked anchorage by the long Detroit Island, which lies across its mouth. The water here is too shallow for sailing, except for vessels of light draft.

Around the shores of this beautiful harbor the clearings come down to the water's edge, and are dotted with substantial summer hotels, cottages and farm buildings. But the traveler on the large passenger boats sees none of these. He passes the island on the west, north or east and sees only frowning shores, crowned with the primeval woods, apparently guardians of Nature's undisturbed solitude.

Washington Island (as has been told in another chapter) was a favorite place of abode for the Indians. Nowhere else in the state are to be found so many of their village sites, cemeteries, mounds, and cornfields. There is here such a wealth of Indian remains that, as one archaeologist says, "there is little left to desire." The entire shore line around Detroit Harbor shows remains of village sites. So also do the shores of Little Lake and Jackson Harbor. Even at this late date a very well defined Indian cornfield can be seen in the grove of timber adjoining the Washington Harbor school on the north. Judged by the amount of Indian remains, Washington Island was the most favored region in the western country.

In one respect Washington Island is still the most favored spot in the Middle West. It has around it the richest fishing grounds on all the Great Lakes. Millions of dollars of fish have been taken from the waters around it, and the end is not yet. At present most of the fishing is done by gasoline and steam boats carrying a crew of five to ten men. They go twenty-five to thirty miles into the lake to set their nets. These are known as gill nets, because the fish in trying to swim through them are caught by their gills. The nets are located by long distance ranges from Manitou Island, headlands on the Michigan shore, and other points. As these nets are set on the lake bottom, sometimes eighty to ninety fathoms deep, they are very heavy. They are therefore reeled in by machinery as the boat moves slowly along the line of nets. As the fish are pulled from the nets they are thrown on the cleaning table where men quickly dress them. By the time the nets are pulled up they are all placed in their boxes and sprinkled with washing powder. A steam jet is turned on them and they are ready for setting again.

Frequently the nets are torn by clinkers which have been thrown from passing steamboats. They also fill up with moss and seaweed. Because of these difficulties, hooks and lines are frequently substituted for nets. They are anchored to the sea bottom in the same manner as the nets. A line, perhaps a quarter of a mile long, is set out horizontally. On this line at intervals of six feet are suspended vertical lines about four feet long. These lines are each fitted with a hook on which a small herring is put for bait. When the trout or whitefish hungrily seizes the herring, he suddenly finds himself caught on the hook.

The early fishermen did not have such elaborate outfits, but the methods used were the same. Gill nets were used to catch chub, whitefish, and trout. Pound nets were used to catch herring, and they are still used in the same way. These pound nets are very large nets set vertically. They are supported by "pound-sticks," from twenty to seventy feet long, driven into the lake bottom. The nets consist of a "lead" extending from the shore to the "pound," which is placed from five to fifteen hundred feet out in water perhaps sixty-five feet deep. The herring in swimming along the shore are stopped by the "lead." They turn and follow it, as they can get neither over or under it. When they reach the "pound," or "pot," they are caught in a trap something like a modern fly trap. It has a meshed bottom, and some-times three or four tons of herring are caught in a single "pound." These "pounds" are then lifted and the fish scooped into a large, flat bot-tomed "pound boat." When the lake bottom has the right slope and depth, one pound net after another is set in a straight line for a distance of a mile or more. On top of the many pound sticks which project from the water a few feet, there is always perched a full quota of sea gulls watchfully waiting for an unwary herring to come too close to the surface.

Almost a hundred years ago the first white men settled on Washington Island, attracted by its rich fishing. The fish then were very abundant. Whitefish could be seen leaping into the air, and sturgeon were so plentiful that they were often stacked like cordwood on the shore, there being at that time no market for them. Trout were incredibly large. Some time later a record was kept which will illustrate what huge fish were caught. In the spring of 1860 Joseph Cornell caught a seventy pound trout off Rock Island. In 1862 William Cornell, a fourteen year-old boy, caught seven trout, the smallest weighing forty, the largest forty-eight pounds. In the spring of 1882 two trout were caught on Fisherman Shoal weighing fifty-eight and sixty-five pounds. They were sometimes just as numerous as they were large. In 1869 Godfrey Nelson caught two hundred and twenty trout in two days. In the winter of 1875 Charles Sloop caught one hundred and twenty in one day, and one hundred and forty the next. Sometimes it required perseverance, but the results were usually satisfactory, as was the case with Silas Wright, who fished for eleven days without a bite and then caught boat loads on the twelfth. These are all hook-and-line catches of authentic record.

With such generous returns for the labor expended there was the usual extravagance which goes with easy money. To make up for the restrictions in the life and diet imposed upon them by their surroundings, the fishermen were lavish in their expenditures whenever an opportunity presented itself. A dollar was a very small coin in those days. Canned goods, fancy toys, laces, and costly furnishings were imported in reckless quantities. Ranney, their easy going merchant and fish buyer, was also their banker, and handed out liberal quantities of cash without any formality of notes or securities.

Nor was there any lack of merry making. As most of the fishermen engaged a number of girls to help them in overhauling and "taking up" the nets, and in hanging them on reels to dry, a "shin dig," or dance, could be arranged at a moment's notice. On special holidays, like the Fourth of July, there was, of course, much boisterous celebrating. A schooner or tug would be hired to take a crowd to Escanaba for grand doings. Another crowd would secure a rival boat, whereupon there would be a race with noisy shouting and laughter. On such occasions drunkenness was, of course, common, and fights would start and end in two seconds.

These hardy pioneers of the deep for many years constituted a sort of fisherman's aristocracy, who looked with pity upon the poor fellows coming in as wood choppers and farmers. They esteemed the land of little or no value except to supply the potatoes they needed to mix with their finny diet. Their thoughts and plans were of the sea, and its vagaries were a constant subject of conversation with them. The land was dull and dusty, but the sea was fresh, and full of riches, sparkling with sport, and full of thrilling adventures.

But that big rolling sea that surrounded them and fed them was also a grim taker of tolls. Many a family on the island mourned one or more of its members who had perished in its treacherous depths. Sudden storms were common, often the greatest skill was in vain, and a widow with her little ones were left to stare disconsolately out yonder where husband and father had gone and never returned. More often, however, these tragedies were caused by the general love for strong drink. There was a boat builder on the island by the name of Bill Stahl. His boats were fast sailers, but they soon got a bad reputation for killing fishermen. A Stahl boat and a bottle of whisky were a combination which was soon looked upon as a sure end for the owner.

The best fishing was often found near the long low shore north of Menominee. Frequently fishermen from Washington Island would cross over to the "west shore" and try their luck at fishing in the fall, returning with their equipment and part of their profits shortly before Christmas, to lay up for the winter.

One fall about 1860 there were among others, three fishermen on the west shore. They were

Ingham Kinsey and Bill Stahl from Washington Island, and Allen Bradley, with his boy, from Hedgehog Harbor. They used gill nets, one hundred eighty feet long, set in from six to ten feet of water. The whitefish were so numerous that fall that they had all they could do to empty the nets. Two hundred whitefish, weighing from three to six pounds each, were frequently taken from a net at each lift.

Pleased with their success, they kept on with their fishing until rather late in December. Finally the day came when they decided to pull up and return home. They loaded their nets and winter supplies into their boats and set sail.

The day was a cold and cloudy one, with a rather steady wind from the southeast which promised to land them on Washington Island in reasonable time. It was their plan to keep the boats together, for Kinsey and Stahl were alone in their respective boats, which fact rendered their journey somewhat difficult. For a considerable time they got along very well. Finally it began to grow dark, and the wind began to swing around to the northeast, blowing briskly dead ahead. It soon veered to the north, and blew furiously, while the weather became intensely cold, the mercury falling almost to zero. By this time the boats had become separated and lost to the sight of each other, and each man struggled as best he could.

But it was a desperate and useless struggle. The flying spray had saturated their clothing, and every outer garment became frozen. Their sails also became stiff and unmanageable, and their ropes like rods of steel. Meanwhile the wind was howling, the waves roaring, while the storm tossed their clumsy craft as it would. They felt the numbness of intense cold and despair coming over them. Through the darkness of the night they were driven helplessly to their doom.

Allen Bradley's boat had outdistanced the others, since it had two men to navigate it. Bradley was, moreover, a very strong man with the endurance of a wild animal. In the coldest weather he was never known to wear coat, over-coat, or mittens. As he sat in his ice encased garments, gradually feeling his limbs turn to the numbness of death, his ear suddenly detected a sound different from the roar of the storm. It was the booming of the sea on the rock-bound shore of Door County. With sudden life, he jumped to the mast and with a tremendous wrench tore it out of the socket. Another jerk or two and the spar and the foresail were also thrown overboard. He then seized the oars, and, seconded by the feeble but earnest efforts of his son, got the heavy laden, ice-encrusted boat under control. Tugging incessantly at the oars, he managed to keep it clear of the shore. After two hours of this work he was finally rewarded by turning the point of a little cove and finding himself safe in Fish Creek.

There was great surprise in the little village the next morning when it was learned that Allen Bradley had arrived during the night. The cold had been so intense and the gale so terrific that it seemed incredible that anyone could have survived it in an open boat. It was generally agreed that the other two fishermen must have perished, but some efforts were made along the shore to discover their bodies. A heavy fall of snow, however, had covered everything with a cloak of white.

Toward evening a searching party from Fish Creek saw a slowly moving body about a mile away. At first they thought it was a bear because it was moving on four legs. They approached nearer and, to their surprise, saw that it was a man moving painfully through the snow on his hands and knees. It was Ingham Kinsey. During the preceding night he had been hurled almost insensible with cold on the beach four miles south of the village. His boat had been smashed on the rocks. During the night and the next day he had staggered along the beach, first north and then south, vainly looking fora human habitation in the unsettled wilderness. Finally his limbs refused to support him and with the last fragment of endurance, he was crawling along, his hands and feet frozen, when he was discovered and saved.

Meanwhile, where was Bill Stahl?

William Stahl was a famous water dog who had survived so many adventures that he believed himself immune from death in the water. He had built both his own boat and the one that Kinsey had, and had unbounded faith in them. Yet he recalled now that his boats had a bad reputation His thoughts wandered back to the long list of fishermen who had lost their lives in boats built by him. There was old Peter Bridegroom who went down the first time he had sailed his boat. Then there were Robert Kennedy, James Love, and Frank Wolf, splendid fellows all, but a little too fond of the whiskey. Were there more? Yes, to be sure. There was Ed Weaver and that fellow Casper, both of whom still owed for their boats. A Bill Stahl boat and a bottle of whisky had sealed their fate. Was the combination going to prove true with him also?

As he felt his boat settling deeper and deeper with its load of ice, and becoming quite unmanageable, he gave up all attempts at navigating her, and devoted his energies to keeping his hands and feet from freezing. But it was a practically useless effort. He was soaked with water and frozen with ice, and more ice was forming around him.

As he listened to the howling of the wind, the swish of the whitecaps, and the heavy thud of a wave striking his bow in the trough of the sea, it seemed as if the resistless cavalry of hell were hitched to his boat, dragging it onward to that brink where he would tumble over into the next world.

Sitting thus, with distressing fancies flitting through his mind, his boat suddenly struck hard on a rock. Before he realized what had happened, another wave followed, throwing the boat upon a rocky beach, while he was thrown into the water.

He scrambled out and looked around him, but nothing could be seen in the darkness. But, now that he was on firm ground he felt new hope within him. He would strike out at once following the shore till he came to a boathouse or human habitation. He stumbled over the driftwood that littered the shore, slipped on the stones, but struggled on. He felt that his limbs were not yet frozen and with good luck he would yet reach a shelter. Suddenly he stopped in amazement.

There in front of him was another overturned boat lying in exactly the same position as his.

He reached into the bow of the boat and pulled out an oblong box. It was his boat. Here was his tool box.

He stared vacantly at the boat. How could it be his boat? He must have turned in his tracks and retraced his steps. Was he losing his mind?

He started along the shore once more, keeping the water on his left, the land on his right. He walked carefully to avoid confusion. At the end of a half hour he was again in front of the boat!

Suddenly he realized the situation. He was on an island, and the reason that he had come twice upon the boat was that he had twice walked around the island.

By this time it was beginning to grow light in the east. By looking in that direction he could now distinguish the high cliffs of Door County. Straining his eyes to the northward he could also discern a long low shore which must be Chambers Island. He now recognized where he was. He was cast ashore on Hat Island, a barren little rock supporting a few stunted trees about five miles southwest of Fish Creek.

The dejection that followed upon this discovery struck him like a blow. He felt excessively weary. He had toiled and struggled and suffered all through the day and night before, his clothing was frozen stiff, he felt numb with the cold. He wanted to sink down and forget it all in a moment of slumber. Suddenly he started up. In his pocket was a match box. There were birch trees on the island. He needed but a little of this bark and in a few moments he would have a fire.

He stripped some bark from the birches, broke some dry twigs from the trees, and struck a match. It refused to ignite. He struck another and another till the box was half empty. Still no success. Then he examined the matches, and found that they were all water soaked. Still clinging to hope, he struck the matches with greater care than before till the last was tried in vain.

Almost stunned by this experience, he went through his pockets one after another. Some were so frozen that he had to tear them apart with main force. What chance had he unless he could make a fire? In his exhausted condition he could not endure it another hour. In his hip pocket was his can of tobacco. It had a tight fitting cover. He opened it and found it almost empty of tobacco. But down there among the crumbs lay a match. He could see no sign of moisture inside the can. He poured out a little tobacco in his hand and

examined it. It seemed as dry as ever. If the tobacco was dry it was likely that the match was
also. He carefully refrained from touching it, however, lest his clumsy fingers might drop it in the
snow.

One humble, forgotten match. Yet it might
mean another lifetime for him.

With this thought in mind he once more made
his preparations for a fire. This time, however, he
proceeded with much greater care than before.
First he picked out the spot on the island which
seemed most sheltered from the wind. Then he
made a windbreak of his frozen sail which he
propped up with a number of supports. Then he
gathered a good sized pile of dry twigs and birch
bark. Finally he carefully gathered up a lot of
dry leaves which he found inside a hollow log.
These he tested for dryness, one by one, before he
put them in their place. Finally he selected a
flat, dry stone, not too rough, under the same log.
This was to strike his match on.

He took out the match, but hesitated to strike.
What if it were defective? What if it broke and
fell into the snow? What if the tinder refused to
ignite? No more boats would sail the bay till next
spring. No travelers would pass on the ice for at
least a month. He knew that within an hour or
two his stiff limbs would be frozen. On the outcome of that match hung life or death.

Lifting the match in silent supplication to
heaven, he scratched the stone gently. It failed to
spark. He felt the sweat break out beneath his

sodden garments. Then he pulled himself together with a jerk, muttered an oath and struck the match with greater force. The flame burst out and he thrust it down among the dry leaves. Then followed an interminable interval. Finally the thin veil of white smoke was succeeded by a leaping flame. Carefully he fed the fire with birch bark, twigs and sticks until he soon had a large fire blazing. He was saved.

Toward evening it occurred to him that he ought to have his fire out on the beach where it might attract attention from the mainland. He made a roaring bonfire, fed with stumps and logs, snatching a nap intermittently.

This fire was soon seen from Fish Creek. When the good people of the village found Kinsey that evening, they appointed a lookout to patrol the beach and keep a watch for Stahl. About midnight the village was electrified into life by hearing this lookout shout in the street:

"Bill Stahl on Hat Island! Bonfire blazing!"

Quickly a willing crowd gathered at the pier to lend a hand in the rescue. But how were they to launch a boat? The heavy gale had packed a sheet of anchor ice into the harbor a mile deep. This had frozen together into a solid mass in many places two feet thick. Meanwhile the storm was still roaring and it needed a good vessel to weather the seas. Saws, picks and axes were found and the whole village went to work to cut a channel. By nine o'clock the next morning a channel was cut a mile long and the best vessel in port was

towed out. Sails were bent and before noon Bill
Stahl saw sweeping down on him a white winged
carrier of life.

The nearest post office and base of supplies
to Washington Island in the forties and fifties
was Green Bay a hundred miles distant. In summer time this did not seem much of an inconvenience for there were plenty of sailboats. But
in winter when the waters of Green Bay froze up
it was a real hardship. It was then necessary to
select one of their number to make the trip to
Green Bay on foot with a handsled. This journey
had to be made on the ice, for the long peninsula
was then a trackless waste of dense timber
throughout its entire length. H. D. Miner was
the best man for this difficult task. Other men
tried it and perished on the windswept, treacherous
ice, but Miner safely brought his sleigh-load of
mail and supplies up and down the bay for thirteen
years. His hardships and adventures in dragging
his sled for two hundred miles was something
unique. Through storms and bitter cold his
journey lay. Sometimes the ice was piled up
edgewise in sharp ridges, and again he came to
gaping cracks and treacherous thin ice where the
invisible currents had melted the ice from below.
But his dexterity and endurance were quite equal
to the task. With good luck he could make the
trip in a week. On the first day he planned to
reach Ole Larson on Eagle Island or Increase
Claflin at Fish Creek, a distance of about thirty

miles. On the second he pushed on to Little
Sturgeon Bay and found shelter with Robert
Stephenson, another thirty miles. The third day
he had the longest march, forty miles, and reached
Green Bay. The fourth day was spent in making
the necessary purchases and then came the home
run with a loaded pack sled. For such a trip he
received up to twenty dollars. But many a time
was his plight so desperate amid the dangers
of the ice and the bitter cold that his twenty
dollars and his far distant island home seemed to
fade utterly away.

Miner was the first postmaster of Washington
Island. The first year he held office his salary was
ten dollars. He was an eccentric man of deep
religious convictions with a burdensome outfit of
rules, precepts, habits and other ironclad regula-
tions of life and daily conduct. Among other
oddities it is told that whatever he took in or
harvested, such as honey, fish, vegetables, etc.,
was always divided into three portions, one for
himself, one for his wife and one for his son, Jesse.
If any of them received company, it was obligatory
upon the one who was honored with the visit to
feed the company and the family out of his or
her portion. Sometimes his wife rebelled at the
straight and narrow path that was laid out for her.
Upon such occasions Miner with the best inten-
tions in the world would tie his helpmate to a
chair and then proceed to administer a dose of
physic and a lecture on proper conduct, meaning
by this double application to purge both the flesh

and the spirit of his consort from the evil that beset her.

For thirty years the fishermen ruled Washington Island alone. The land was considered too far north for farming, the woods too formidable, the soil too stony. But in 1868-70 came some groups of stout Norwegians, Danes and Icelanders who secured homestead rights back in the timber. They did not expect to do much in the way of agriculture. Their main hope was in cutting cord wood. This they set to work with great energy to do. Soon the mighty maples swayed and fell and were then split and cut into four foot lengths with an axe, for cross cut saws had not yet come into use. The price, delivered at the pier, was two dollars per cord, an immense amount of toil for a pittance. Yet it was better than nothing. Frequently there was no sale for cordwood and they were then obliged to roll huge logs together and burn them. When a little field was finally cleared the stumps stood immovable for many years, an obstacle to cultivation. Meanwhile they had nothing to sell, and their distress was great. With but small pastures and little hay, their cows dried up in winter and gave no milk.

Unexpected difficulties also developed, chief of which was the difficulty of securing water. The story is told that one of the Danes set to work the first summer to dig a well. He got down only a few inches when he came to a flat rock. He dug and dug on every side to get around this stone, but it seemed to stretch out indefinitely. He told his neighbors about the trouble he was having with it. They came and inspected the difficulty. Then one of them started to dig on the other side of the cabin. Only a few inches down the same stone appeared. Filled with evil forbodings they hurried home and began to dig around their own huts. The same stone appeared also there, for it was the solid rock which underlies all of Washington Island only a foot or two below the surface.

Since they could dig no wells they were obliged to carry water from the lake, in pails, in kegs, on wheelbarrows and in barrels on wagons with oxen. As it was so laborious to get water they were obliged to be very saving with it.

But outside the borders of their island lay the water, a hindrance to communication with the outside world, and especially so in fall and the first half of the winter when navigation ceased and "the Door" had not yet frozen over. They felt like stranded mariners a thousand miles at sea. L. P. Otteson recalls how they once went for seven weeks without word from the outside world. This was bad, but what was worse was that the whole island had run out of chewing tobacco. All possible substitutes were tried, such as willow bark, juniper twigs, cabbage leaves, etc., but without relief, and further abstinence was intolerable. Finally, Mr. Miner consented to go. It was his last trip to Green Bay on the ice. A long and dreary week followed. At last a large party of young fellows walked out on the ice to meet him, or rather the quid, half way. When he

appeared in the distance they broke into a run
and soon were eagerly pulling at the strappings of
his sled. The tobacco was found and immediately
passed around, each one snapping off a generous
allowance with intense relish. There was a minute
of silent bliss, wherein the movement of many
jaws was faintly audible. Then they all turned
homeward, staining the ice an odorous brown and
feeling that all was well with the world.

Undismayed by the many obstacles of nature
these Scandinavians stuck to their task and, in
spite of all evil prophecies, turned this formerly
unproductive island into beautiful farming land.
The stumps were blasted with dynamite, the deep,
dark woods were turned into sunny fields, and
well drilling machinery was found which pierced
that solid layer of limestone a hundred feet deep
and found waterbearing strata beneath. Even the
inumerable rock fragments which everywhere
littered the ground were finally turned into good
use, because they were crushed into first class road
material, giving the island smooth and excellent
highways.

Washington Island now exports thousands of
tons of foodstuffs annually, such as potatoes,
butter, grain, fish and fruit of all northern kinds.
Few places in America have such a diversity of
export products as this island. It is as progressive
and enlightened a community as can be found
anywhere.

CHAPTER VIII

A MAN OF IRON: A TALE OF DEATH'S DOOR

He was a man; take him for all in all,
I shall not look upon his like again.
                                        SHAKSPEARE.

Port des Morts the French called it; the Door of
Death. This was not an invention of the French,
for the Indians had learned its treacherous nature
and so called it long before the white men came to
visit them. The French merely translated the
Indian name into their own tongue after learning
by personal experience how well it applied.

It is not long, this door of death, nor wide. It is
merely a passage between the tip end of the Door
County peninsula and the islands beyond. But
in this strait were often met strong currents and
fierce winds running counter to each other, which
baffled the seaman's skill and drove his craft on
the rockbound shores. Here it was, according to
tradition, where La Salle's *Griffin*, the first vessel
to sail the waters of Lake Michigan, met her doom
in 1680, a fortune in rare hides going down with
her. Since then hundreds of vessels have here been
flung ashore and wrecked. One week in September,
1872, no less than eight large vessels were wrecked
or stranded here. In the summer of 1871 almost a
hundred vessels suffered shipwreck in "the Door."

Just as turbulent as are these straits in summer,
just as treacherous are they in winter. The ice

105

forms late and breaks up early. At no time is it entirely safe. Shifting currents undermine the ice without ceasing. Where the ice may be several feet thick in the morning the waves may wash in the evening.

Many stories could be told of terrible adventures in crossing this treacherous bridge of ice. Many a man and horse have had a desperate battle with death while plunging through the perilous ice, and more than one man has seen his last hope of life perish as, clinging to a cake of ice, he has been driven out into Lake Michigan where soon his frail craft would break up.

As an illustration let us recount the story of Robert Noble's experience in "the Door," not because it was the worst, but because he lived to tell of his icy battle. Also because in human endurance it is almost unique among the tales of suffering.

On December 30, 1863, Robert Noble left Washington Island after having spent some happy Christmas holidays in visiting the girl of his heart. He was a splendid young fellow physically, was twenty-five years old, weighed two hundred and twenty pounds and stood more than six feet in his stockings. He had a flat-bottomed skiff, and for a while had little difficulty in making his way among the broken ice floes in "the Door." Abreast of Plum Island, however, he struck a large field of thick ice through which it was impossible to force a passage. With some difficulty he finally made a landing on Plum Island, hoping that the wind might clear a passing for him to the mainland a couple of miles away. It was now getting dark, snow had begun to fall, and the weather which had been mild was getting very cold.

While groping about on the shore of the deserted island, he finally came to an abandoned fishing hut which had neither roof, doors, nor windows. Here he made a fire, but had difficulty in keeping it burning because of the falling snow. Toward morning it went out altogether. Ice was now forming all around the island. When he saw that he would have to remain there for some time, he thoroughly explored the island. He found that the only other building was a ruined lighthouse, of which only the cellar and chimney remained. Here was a sort of a fireplace and he managed after much trouble to light a fire, but not before his last match had been used. He heaped this fire with such fuel as he could find, and became more hopeful as its warmth began to be felt. But suddenly his hopes were blasted. The chimney was full of snow. It began to melt, and soon there was a rush and tumble and his fire was buried under a heap of snow. It was a most depressing blow, since it was now getting dark and the weather was becoming bitterly cold.

He had a revolver with him and made a number of attempts to start a fire by putting strips of lining from his coat over the muzzle, hoping that the explosion would cause the cloth to catch fire. But this was all in vain. Yet he managed to hold out in the little cellar all night without food, sleep or heat. Through the interminably long hours of that bitter night he paced about in his prison, keeping from utterly freezing by all kinds of exercise, moving stones and logs about and otherwise exerting himself. Finally the gray dawn of January 1, 1864, appeared.

January 1, 1864! Old settlers have not yet, after a lapse of sixty years, forgotten the intense cold of that day. Tales are told of water freezing by the side of heated stoves,·of the impossibility of keeping warm in snug beds, of cattle freezing to death in their stalls. It is remembered as the coldest day in the history of Door County.

Robert Noble did not know anything about this. He only knew that it was fearfully cold, that he was starving and that he had gone for two nights without sleep. He realized that his only hope was to leave that deserted island at once. The wind had now broken up the ice which was bobbing about in a slushy formation. He launched his boat and for a quarter of a mile managed to row his boat toward Washington Island. Then he encountered solid ice and could make no further progress with the boat.

As the ice was not very thick he tore out the seats of the skiff, and by help of some ropes fastened them to his feet in the shape of rude snowshoes. He hoped in this way to distribute his weight on the fragile ice. For a few steps this worked satisfactorily, when suddenly the ice broke and he was plunged into the water. Fortunately he had a long pole with him, which saved him from going under the ice. He tried to kick the boards from off his feet, but could not. By hanging to the pole with one hand he managed to secure his pocket knife and, reaching down, cut the ropes that held the boards to his feet. Finally he managed to get out of the water and back to the boat.

He was now extremely cold, his wet clothing had frozen to his body, and his arms and legs were encased in an armor of ice. Yet such was his splendid vitality that by stamping and tramping about in the boat he once more got circulation through his limbs. As soon as this was obtained, again he took his two boards and, lying down on them so as to distribute his weight over as large a surface as possible, he attempted to pull himself, snake fashion, toward the shore of Detroit Island, about one and a half miles away. He had not gone far, however, before the ice broke again and he went down head first. By the time he could turn over in the water the current had carried him under the ice. Then followed a terrible struggle, hampered as he was by his heavy and frozen garments. In youth he had accustomed himself to diving and remaining under water a long time. This now saved his life. After an interminable struggle against the current and the ice he finally regained the surface through the hole he had fallen into.

He now gave up the attempt of gliding over the treacherous ice by means of boards or otherwise. Instead of that, he stayed in the freezing water, using his ice-encased arms and hands as sledge-hammers to smash the thin ice and open a passage. He slowly moved forward, like an animated ice-berg, half swimming, half crawling, by help of his elbows. When he came to a floe of heavy ice he pulled himself on top.

This incredible struggle against the merciless elements continued for hours. Time after time he believed himself lost, but again and again he conquered, smashing, plunging, rolling and swim-ming, with the temperature at forty degrees below zero.

Thus he continued until late in the afternoon, when he reached the shore of Detroit Island. Here he encountered a high barrier of ice made by the freezing spray of the waves. Loaded down as he was with such a burden of ice, he was not able to pull himself over this obstacle. Finally he found a tunnel in the barrier, such as is sometimes formed by a spiral of spray, and wormed his way through it to the land.

By this time his feet and hands were frozen and senseless, but yet he was able to keep on his feet. He crossed the ice of Detroit Harbor without further accident, and came about dark to the house of a fisherman. He was met in the door by the owner who stared amazed at this bulky ap-parition of ice in the shape of a man. Quickly Noble explained what had happened and begged him to provide a tub of water in which he could put his feet and two pails for his hands. This was done and immediately the poor sufferer, who had

had no food or sleep for three days and two nights, fell asleep.

Unfortunately for him a meddler just then appeared upon the scene. A neighbor came in who insisted that cold water would not help. He told of kerosene, the new mineral oil, a shipment of which had shortly before reached the island, and of which exaggerated stories were in circulation. This, he claimed, would take the frost out. The kerosene was found and the poor man's hands and feet were soaked in this oil. But the kerosene was bitterly cold, far below the freezing point of water, and, instead of taking the frost out, it effectually prevented the frost from leaving the affected parts. When Noble awoke he found his limbs were frozen beyond remedy.

Then followed bitter months of suffering for poor Noble. There was no physician on the island. The nearest was at Green Bay, a hundred miles distant. Nor was there any means of getting him there. There was not a horse or an ox on the island, and most of the able bodied men were off to southern battlefields. Bert Ranney, the Washington Harbor storekeeper, ever ready to help a sufferer, took Noble to his home and gave him as good care as possible. Here for month after month Noble sat, as helpless as a child, enduring agonies of pain, and in dreary idleness. One by one his fingers dropped off and little by little the flesh of his legs peeled away. After a while only the white, lifeless bones of his feet were left, while his system with never ceasing pain and agony strangely adapted itself to the changing conditions.

Finally, in June, 1864, an opportunity presented itself to send Noble away. A Doctor Farr, from Kenosha, was temporarily in Sturgeon Bay while negotiating the purchase of a saw mill. He was willing to perform the needed operations, but he lacked the necessary surgical instruments. He obtained some from Green Bay; but the only saw available was an ordinary butcher's saw. With this rough tool Noble's legs were amputated below the knees.

The operation was successful and in due time Noble once more felt able to work. By the help of friends he obtained artificial limbs, and soon was back at his business of drilling wells. In spite of his lack of fingers, he developed a remarkable dexterity in handling the tools of his trade, and he was never one to ask for favors because of his physical handicap. Later on he operated a ferry between Sturgeon Bay and Sawyer for many years.

Such energy, such extraordinary endurance, such fortitude in suffering should be rewarded with a public monument, and a pension. But unfortunately the keen competition of the later years drove this sturdy pioneer to the wall, and his reward was finally a "home" in the poorhouse.

CHAPTER IX

"THE MARRIED BUCK"

A STORY OF CHAMBERS ISLAND

Full thirty feet she towered from waterline to rail,
It cost a watch to steer her, and a week to shorten sail;
But, spite all modern notions, I found her first and best—
The only certain packet for the Islands of the Blest.
                                                    KIPLING.

The largest island in Green Bay is Chambers Island. It is about eleven miles in circumference and its straight shore lines do not present an inviting appearance to the passing traveler, except upon the northwest side, where a large bay indents the land. About a quarter of a mile from the head of the bay lies a beautiful lake about a mile long and half a mile wide. Two small timbered islands with reedy shores are in the lake. The lake is separated from the east shore of the island by a high ridge of gravel only a few feet wide, but well screened by a sturdy growth of pines.

Chambers Island received its name in 1816. In August of that year, Colonel John Miller was sent to the spot where the city of Green Bay now lies to establish a military post. While the territory embraced in the state of Wisconsin had been ceded to the United States by the treaty of 1783, England did not relinquish her hold upon it until July, 1815. Now Colonel Miller with five hundred men were sent to take actual possession of this

113

distant western region. The military force accom
panying him, with their considerable supplies, were
conveyed to their destination in four large sailing
vessels. This party named many of the localities
visited by them, and these names still remain.
The largest island in the bay they called Washington Island, in honor of the father of his country,
and also because the flagship of the fleet was
named Washington. The next largest was named
Chambers Island in honor of one of the officers of
the expedition.

To most people Chambers Island is but vaguely
known as a desert island lying far out amid the
waters of Green Bay, and the largest private
estate in Door County, where the deer roam unmolested through the forest arches. There was a
time, however, about fifty or sixty years ago,
when Chambers Island was not only a settled
community, but was an organized town, where
the voters once a year gathered in solemn conclave
to discuss the needs of roads and bridges, distribute
political honors, and grumble at the state tax. It
had a full array of town officers, with no less than
three justices of the peace, three constables, and
even a "Sealer of Weights and Measures." It had
a public school and a postoffice, farms, orchards,
livestock, a sawmill, and a ship-building plant.
It had occasional religious services and a sewing
circle, and tea parties met frequently. The isolated
little island was in all respects a well ordered
community.

Eventually their work, which was chiefly that
of cutting the timber, came to an end, and the

settlers moved away to other parts. Their schoolhouse and their dwellings fell into decay. So also
did the countless heaps of brush and rubbish, and
Chambers Island is once more reverting to the
sylvan dreamland it was when the troops of 1816
first saw it.

Among the men on this island who long ago
toiled, planned, and dreamed of better things, only
to pass away and be forgotten, was one of such
unusual energy, such indefatigable perseverance,
and such constructive ability under the most unfavorable circumstances, that he holds a unique
position even among the stalwart pioneers of all
regions who have led the way in blazing a path to
civilization. It is to his achievements that this
sketch is dedicated.

Long ago, some time before 1850, there was one
lone settler on Chambers Island. He was a tall,
grizzled Irishman, named Dennis Rafferty, and
he was a fisherman. Where he came from is unknown and immaterial.

A short time after Rafferty settled on Chambers
Island, there came to Green Bay a young sailor
from Norway. He was a keen-eyed, wiry young
fellow about thirty years old. He had about him
an air of dashing fitness such as we are wont to
associate with heroic knights of old, but like them
he had also one serious limitation, he could neither
read nor write. Since he was ten years old, he
spent most of his time following the sea. Now he
was married, and had promised his wife to find
another livelihood. His name was Johannes

Bukken, but when he came to Green Bay he was
soon told that that name would never do. His
employer was another Norwegian, a big, pompous,
fisherman whose name was Jeppe Meraker. This
he had changed into the more euphonious Jess
Morefield. He assured his newly arrived assistant
that he must change his name to John Buck.

"Oh, all right!" said John. "If that is necessary,
let it be. But what about my wife's name; must
that be changed too? Her name is Marit."

"Hm," said the man of experience, "I am afraid
that name will never do."

"Oh, yes it will," replied John, "for a better
name is not to be found in either America or Norway. And Marit shall be her name as long as she
is mine!" He was tempted to add that a better
wife was not to be found either; but Marit with
the sky-blue eyes and the deft, gentle hands,
occupied a sacred place in his heart, and he did not
wish to reveal such tender sentiments to his rather
cynical employer.

One morning John was sent out to look after
some fish nets quite a long distance down the bay.
He had one of the flat bottomed, cumbersome
boats, called pound-boats, which were in general
use by the fishermen on the bay. Just as he arrived
at the nets, a very strong wind sprang up from the
southwest. Before he was through with his work,
the bay was foaming with whitecaps. He took in
the main sail, but let the jib stand, and cast loose.
Then he noticed that he had lost the rudder. It
must have worked loose as the boat was bobbing

up and down against the tightly stretched "lead."
No matter, thought John, running across the seas
in this howling wind was impossible, anyhow. He
thereupon seated himself in the stern, took the jib
sheet in his hand, and let the boat run before the
wind.

It was a glorious trip. By help of the jib he had
no difficulty in holding the boat straight before
the wind. For long intervals the boat would rush
forward, perched on the crest of a big wave, with
the bow hanging over the hollow in front. Finally
the boat would glide gently down into the smooth
trough below, while the greenish white foam vainly
snapped after him like wild dogs. There his speed
slackened until another wave again hoisted him
upon its foam lathered crest. It was like riding
wild horses. It gave him a keen longing for his
old sea life again.

Late in the afternoon he came abreast a large
island. But the waves were breaking with such
force on the beach, that it was impossible to land
without smashing the boat. He kept clear of the
shore until he rounded a high point and saw a large
cove opening into the lea of the island. At the
head of it he saw a dim light from a cabin blinking
in the gathering dusk.

He now concluded that he had drifted far
enough. Moreover, he was ravenously hungry.
He therefore made fast the sheet, kicked loose a
thwart, and with this as a rudder, soon had his
craft landed in front of the cabin.

This was the house of Dennis Rafferty on
Chambers island, and here John was given a

boisterous welcome. He was the first guest that Dennis had had since the hut was built, more than a year before. John stayed there that night, and they sat by the fire talking of fisherman's luck until late in the evening. John soon found that the fishing was much better in this part of the bay than in the shallow waters near Green Bay. Finally it was agreed that John was to move to the island and become a partner of Dennis.

John and his wife soon became pleased with their change of location. It was like divine worship to them to walk through the grand, silent forest that covered the island. Never had an axe touched those hoary giants, and pines and oaks reared their sturdy columns, some of them three feet or more in diameter, and almost two hundred feet high. As John walked about, the ground covered with a carpet of pine needles that deadened every sound, here and there laying an admiring hand upon the bole of some giant perhaps five hundred years old, he had a far deeper feeling of awe than he had experienced when he had occasionally glimpsed some mammoth cathedral in a foreign port. With Rafferty's aid he soon had a comfortable house built of small pine logs so straight and tight-fitting that they needed no chinking to make them snug. Thereupon he and Dennis became very busy with the fall fishing, and Dennis soon learned to have a great respect for the young sailor's skill. In fact, the jolly Irishman soon loved John as if he had been his younger brother, and nothing was too good for him and Marit. However, this did not prevent him from playing whatever

innocent pranks he could on John, which were
usually repaid with interest. When they were not
occupied with fishing they spent their time in
enlarging their little clearing, where they had an
abundance of vegetables. They also kept a flock
of hens. And as the years passed by, the time
passed pleasantly for these pioneers on their
island domain. They were the only settlers between Little Sturgeon and Washington Island,
with the exception of Increase Claflin who lived
near Fish Creek. Marinette and Menominee
were known only as the site of a squalid Indian
village.

A couple of years after John had come to the
island, its population was increased by the arrival
of Dennis Rafferty's mother. Among other things
she brought a small bag of wheat with her from a
small hamlet in Ireland. These kernels of wheat
Dennis was as fond of as if they had been gold.
He immediately brought them over to John to
show them.

"Here is wheat, my boy, which it is worth your
while to look at. This is something quite different
from the chaff they grow in this country. This
wheat my father and my grandfather and their
ancestors have grown ever since the holy fathers of
Shandon Abbey received it by a miracle from
heaven a thousand years ago. Three years they
were without a crop and there was no more seed
wheat left. Then the holy fathers prayed to the
Virgin Mary to relieve the need of the people
and she commanded them to turn the mouth of

the abbey bell upward. They did as they were told, and the next morning found four bushels of wheat in the bell. Half of it they used for the holy wafers and the rest of it for seed. Since then there has always been fine wheat in Ireland. Now you will see how it will grow, and what excellent bread we will have."

He immediately set to work on a new clearing which he burned and cleaned most carefully. Then he made a harrow with heavy wooden teeth which he dragged back and forth between the stumps until he managed to get the soil quite well stirred. Then he sowed his wheat and covered it well by more harrowing. Warm showers came and soon the young wheat stood green and promising.

When midsummer came and the wheat was ripening, Dennis was in constant excitement lest something should happen to it. He shouted at birds and chickens if they approached the field, and swore until he was black in the face if a rabbit dared to hop through it. Finally, he shut up his hens and asked John to do likewise. John promised to do so, but the next morning his chickens were out as usual. Then Rafferty became angry and shouted:

"Can't you keep those pesky hens of yours shut up a few days, John? They will ruin that wheat yet."

"Yes," said John, "I will see that they are shut up. I forgot to tell my wife last night."

"Well, see that you don't forget it again, John, or I will kill every hen that comes near that field."

Now it happened the winter before that Dennis had played one of his little jokes on John. One day he had suddenly come into John's cabin with an old newspaper, wherein he had read aloud the important news that the legislature of the state had just passed a law requiring all immigrants who had not become citizens to present themselves before the nearest sheriff on St. Valentine's day, the 14th of February. Whoever neglected to do so was liable to thirty days in jail.

Of course, nothing like this was printed in the paper. It was only a joke of Rafferty's. But John did not know this, for he could not read English. He naïvely thought that if such were the requirements, it was up to him to meet them. When the time came, he got up early one morning and travelled off on his skis sixty-five miles to Green Bay, where was the nearest sheriff. To this official he presented himself timidly and awkwardly. The sheriff, whose mind dwelt on evil-doers, thought John was a repentant criminal, and took him before the district attorney. Here John was given a searching examination, but as the vocabulary of the prosecutor was new to John, it was rather slow work to get at the supposed criminal's confession. Finally a couple of interpreters were called in and the matter was explained to the amusement of all except John.

Now this foolish trip to town was no serious matter to John, for he needed to go anyway. Still he felt that he owed something to Dennis for it. The morning after Dennis' explosion about the chickens, John got up early. He shut up his chickens as he had promised, and went over to Dennis' chicken coop and opened a window.

After a while Dennis awoke and as usual went to the window to take a glance at his beloved wheat field. How yellow it was becoming! He must surely go to Increase Claflin and borrow his "cradle" so that he might cut it. Then suddenly his joy was changed to bitter wrath: "If there aint those doggoned chickens of John's again. I will fix them, I will—." Anger left him speechless, but he seized his gun and began to shoot chickens with deadly vengeance. Little did he think they were his own. Soon quite a number were dead, while the others, broken-winged and lame, cackled and hopped off to the woods as best they could.

During this episode, John had come out of his house yawning and stretching, and approached the battle field. Dennis poured a flood of abuse upon him. Finally he threw the dead hens in front of John and said with a chilling voice, "Here, take your damn hens. Tomorrow you will get the rest if they show up."

John picked them up quietly and meekly said, "It is too bad, Dennis, that you are bothered this way. I would do the same myself."

It was not until late in the evening that Dennis returned from his trip to Claflin's home. Then he learned from his mother that he had been slaughtering his own chickens. That was the last straw for the honest Irishman. Without waiting to eat, he marched out to have a settlement with John.

Nothing short of a downright licking would atone for this. Nothing less.

In the meantime, John was keeping a lookout from his house. When he saw Dennis coming toward the house with quick steps, he extinguished the candle. Just inside of the door was a trap door leading into the cellar. This he opened and withdrew into a corner to await developments.

Dennis' steps were soon heard on the porch outside. He did not stop to knock, but kicked the door open. He would take the knave right out of his bed, that was what he would. He stepped in, when the floor suddenly disappeared beneath him and he shot down the cellar steps with terrific noise and many bruises. Immediately he heard the shutting of the trap door. Then a heavy trunk or a cupboard was pulled over the opening and all became quiet. In spite of all his shouts, threats and exertions to get out, he could not hear a sound above.

The next morning John called down to Dennis.

"Good morning, Dennis!"

"Good morning, John!"

"Did you go down to the cellar to look for chickens, Dennis?"

No answer.

"Will you promise to be a good boy if I let you out?"

"Yes, John," came the humble answer. "You win this time."

This incident made no break in their friendship. On the contrary, after Dennis had cooled his wrath by sitting on a potato bin all night, he laughed heartily when he came out. That John should get him to shoot his own chickens and then trap him like a rat in a box, answered so fully his own Irish conceptions of humor that he respected John more than ever. But he meditated revenge, nevertheless.

At this time John was occupied with a great undertaking. He had started to build a schooner. He had for some time realized that there was no future in fishing. Although there was an endless abundance of fish, the profits were small on account of low prices. Often it was impossible to sell and much of it had to be thrown away after it was cleaned and salted. To own and sail a trading vessel was something quite different. Settlers were pouring into Green Bay, Milwaukee, and Chicago and the great sawmills were beginning to turn the pine forests into lumber. Ships were needed to carry all this freight, and John decided that there lay his opportunity. When he first mentioned his project to his partner, Dennis was very enthusiastic about it. He thought the idea was simply to build a large sail boat. But when he found John preparing to lay out a huge oak keel 120 feet long, he stared at the shipwright in amazement. "Are you crazy, man? Are you thinking of crossing the Atlantic?" And thereupon he attempted so many jokes about "John's Noah's Ark," that John became quite disgusted and determined to build his vessel alone without any help from

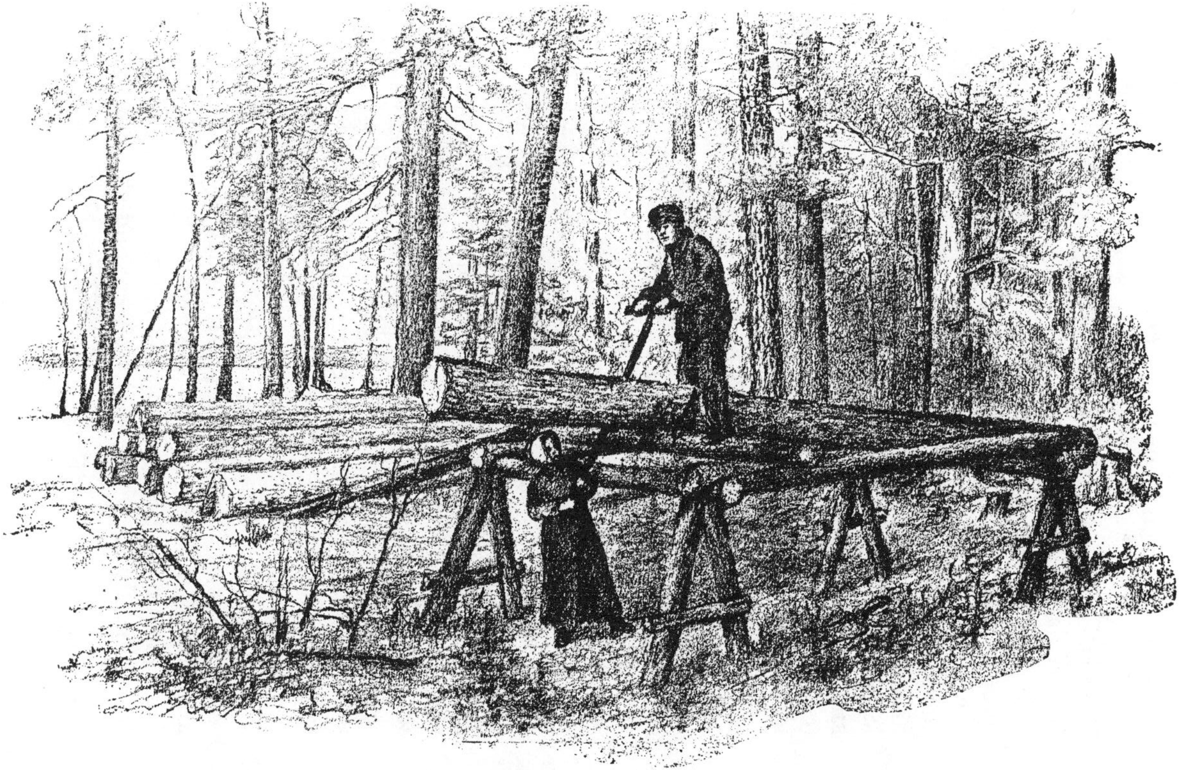

A Pioneer Saw Mill

Dennis. He would show this Irishman what a
Norwegian sailor could do!

John was not as inexperienced in this sort of
undertaking as Dennis thought. He had followed
the sea for twenty years, and was as familiar with
the smallest details of a ship as a farmer is with
his stable. Moreover, he had worked sufficiently
long in the shipyard at Larvik to know the nature
and use of the different tools employed, and the
usual manner in which the ship's timbers are
joined together.

He went to work with quiet resolution. The
massive oaken keel was laid in place on rollers
near the shore. The ribs and deck beams also were
of oak, cut and trimmed with his axe and then
steamed in a box so that they might be bent to the
desired shape. For all other purposes he used pine,
of which there was an abundance on the island.
To be sure, he had no sawmill to cut them up;
but he rolled the logs up on high trestles and then
he and Marit sawed them with a whip saw. Up on
the log stood John, and Marit below, sawing their
way to fame and great achievement through
hundreds of logs. As bolts and nails were expen-
sive, he did not use many of these, but pinned his
vessel together with wooden trummels. They had
the advantage of not rusting, and they made the
vessel lighter. The greatest labor was to raise the
masts. But with block and tackle, much ingenuity
and many failures, but with indomitable persistence, he finally got these also in place and wedged
fast. Finally came the many sheets and shrouds,

blocks and halyards.   Here the work went fast
and also the money.

All this took a long time, more than three years;
but the two, John and Marit, did it all alone, built
the vessel from stem to stern, from keel to truck.
Finally the great day came when she floated in the
harbor, a three masted schooner, shining in her
new paint, with swaying sails which Marit had
sewed. High up on the peak of the main-mast, a
blue pennant with a white star was fluttering,—
this was from Marit's wedding dress. There the
vessel lay, graceful, imposing, fit to battle with
the tumbling seas, carved out of the primeval
forest by two pairs of hands.

Only one thing now remained, and that was the
name.   She was to be called *The Marit Buck*.
Since this name was to be her crowning glory, John
wanted it to be painted by an artist. As this task
was not among his gifts, he induced Dennis to help
him sail the schooner to Green Bay. There it was
to be registered, insured, the name to be painted,
a crew hired, and a cargo obtained.

Arrived at Green Bay, the proper official was
soon found to inspect the vessel and make out
registration and insurance papers. When he was
filling out the blanks he inquired:

"What is the name of the vessel?"

"*The Marit Buck*," answered John in his stiff
English.

"How do you spell the name?"

John scratched his head. To spell in English
was Greek to him.

In the meantime Dennis was ruminating on
what seemed to him a brilliant idea. Ever since
the affair with the chickens he had looked for an
opportunity to repay John for his humiliation, but
no adequate chance had yet presented itself. Now
he suddenly lifted his head with a serious mien
and said: "I will write it for you."

The agent looked at the paper which Dennis
handed him. "*The Married Buck.*" That was a
queer name. But thus it had sounded from the
owner's mouth. The officer filled it in and said:

"It is well that you came this morning, because
I am off for Milwaukee and will not return for
two weeks."

As John had much to do in selecting a crew and
find a cargo, Dennis undertook to buy the provisions, dishes, and cooking utensils, and to find an
expert painter.

Toward evening John came back in splendid
spirits; a cargo of wheat for Buffalo had been
secured and most of his crew had been found. He
needed only a dependable mate who could also
help him with the accounts. He would rather have
Dennis for this, but Dennis was unwilling to leave
his mother.

As he approached the dock, he saw quite a
crowd of men gathered on the pier and looking
at the schooner. John's heart swelled with pride.
They had reason for admiration, he thought, for
a trimmer schooner had not been seen in the harbor. And he had built her from the flying jib to
the spanker boom. Tomorrow she would be filled

with wheat, and then off to Buffalo with himself as captain. Captain Buck! It had a fine sound. It was the proudest moment of his life.

But, alas, it is truly said that pride goeth before a fall. When he reached the pier he saw among the crowd his old acquaintance, Jess Morefield, pompous and fat as ever, and asked him with ill-concealed exultation, "Well, what do you think of the schooner?"

"Oh, I guess the schooner is all right," answered Jess, "but it is that queer name we are looking at." He then pointed to the bow where in large, richly ornamented letters was painted

The Married Buck

"Why, what is wrong with that?" inquired John. "That is my wife's name."

A boisterous laugh of derision broke out around him. "Ha, ha, ha!" Even the pompous Jess Morefield lost his dignity in the abandonment of merriment, and his huge sides heaved with laughter. But he recovered his gravity for a moment and said:

"No, my boy, I don't think that is your wife's name. Don't you understand that the *Married Buck* means 'the buck who has found a mate'? Of course, I understand you're tickled because you found a wife; but that you are so silly about it as to advertise it on the bow of your schooner, that is too much." With this Mr. Morefield once more let dignity go to the winds and roared with merriment.

John turned on his heel in disgust and met the grinning Dennis.

"Now, you rascal, what have you done? You must have spelled the name wrong!"

"Spelled the name wrong!" answered Dennis. "I spelled it as you spoke it. Isn't it right?"

"You know very well it isn't right, Dennis," said John sadly. "To think that you, my best friend, would make a laughing-stock of me like this."

"Hm," said Dennis thoughtfully. "I guess you are right. It was a fool trick. I wonder if I got a little rattled in the head when I fell down those cellar steps," he added with a sly side glance at John.

John walked off.

"Where are you going, John?"

"To find a painter who can fix this mess."

"But that won't do, John. There is a big penalty for changing the name of a registered vessel. Moreover, if you were to have a shipwreck or fire you couldn't collect any insurance money if the name of the vessel is changed. The law is very strict about that."

"But this wretched name, I can't stand it."

"Oh, let it go for a year, John; then we can fix it when you renew the papers. And if you will forgive me for this, I will sail with you and help you as well as I can in writing and in business accounts. You will need that if you don't want to get fooled worse."

A few days later the local weekly had a long account of the new vessel which had been built by a single man without bolts or irons. It was the first vessel built on the bay and it was greeted as a great event. Dennis had given the editor a long account of John's persistence and ingenuity, and told how he had literally sawed and planed his vessel out of the primeval forest. This account was reprinted by other papers and John at once became a famous man. But he did not know this, for when it was printed he was far out on Lake Michigan, absorbed in the study of elementary business methods.

It was a dark September night and a schooner was drifting with reefed sails in a great storm on Lake Erie. It was John and his schooner. For almost a year he had sailed the Great Lakes and had become an experienced skipper. He had visited many ports and the schooner with the queer name had attracted much attention. Now he was on his way from Cleveland to Port Huron with sand ballast to take on a cargo of salt. But evil luck struck him. Outside of Belle Island he was overtaken by a furious gale. As the vessel was running very light, it was a mere plaything for the waves, and John saw himself drifting helplessly toward the shore where the waves were dashed to flying spray. He felt the vessel waver in the shifting current of the undertow and expected to be instantly crushed against the rocky shore. To his amazement, just as the crash was to come, the vessel was lifted by an unusually big wave and thrown like an eggshell clear over the stone beach and into a marsh beyond.

Dazed by the tremendous shock, and with many bruises, the crew crawled out to take an inventory of their injuries. Fortunately, not one had been seriously injured.

But the schooner? There she lay on her side in the slimy ooze, her rigging entangled in the brush, safe, to be sure, from the fury of the storm, but apparently doomed never again to ride the waves.

That at least was the opinion of the insurance company, whose nearest agent after inspecting her reported her a total wreck, and John at once received the insurance money.

John sat on the bowsprit and ruefully inspected the plight of his schooner. In retrospect, he saw the sturdy oaks and tall swaying pines on Chambers Island. He thought of how he and Marit with ceaseless energy and hard labor had felled those ancient giants, and with axe and saw, plane and augur, had changed them into a ship. That vessel was to be a monument to all his wife's virtues, to her cheerful fortitude in the wilderness, to her hopeful endurance in toil, to her youthful grace and her abiding love. He loved that vessel like a home and fireside, yes, like a part of his own flesh, partly because it was his great accomplishment, but mostly because Marit's share in its construction was to him like a constant benediction.

The planks of the deck that he walked on she had helped to saw, the bulging sails were her handi- work, the waving pennant high up at the top of the mainmast spoke to him of her love and the children at home. Probably she was at this moment sitting in the lonely island cabin telling the children that father was on his way home in father's and mother's own schooner. And here lay the schooner half submerged in the slime of a swamp, her deck smeared with black mire. Was this the end? Was she to lie there, a hive for hedgehogs and water snakes, soon to be covered with green moss and trailing creepers? No, never. Let her meet her end, if need be, in the floods which were her element, but he would never suffer his vessel to sink out of sight in the mud of a nameless swamp. To Marit's honor he would make her float again.

There was a shallow bayou or slough in the swamp where the vessel had been tossed. The captain found that after it had meandered through the swamp for some distance, the slough com- municated with Lake Erie, about a half mile away. At that point it was obstructed by a broad gravel bar rising above the lake level. The captain decided, however, that if he could get his vessel to that point he could cut a channel through the bar.

The problem was how to get his vessel to the bar. There was a little water in the bayou, but quite insufficient to float the schooner. But John set to work with his usual determination. First, he purchased the vessel from the insurance company for an insignificant amount. Then he procured several large kedge-anchors and also some big blocks and tackle. With the help of his crew of six men and some nearby Indians, the kedge-anchors were securely fastened a couple of hundred feet in front of the schooner. The pulleys were fastened to these and the ropes then carried through them back to the windlass on the deck. The entire crew then manned the windlass and the vessel was heaved forward inch by inch through the slippery muck. Some days they could make a hundred feet or more, while on others, it took hours of work to budge her a foot. But John always found some way to move her forward. Finally, after a month's toil, he had the vessel nosing against the bar. To dig a channel through the reef was easy. Then the captain waited for an east wind to raise the water. It came and the vessel glided smoothly out. Its injuries were slight.

The first thing John did when he was once more under sail, was to bring out a box full of large shiny brass letters. "Now that the schooner is born again," he said, "she ought to have a new name." Then he laid the letters out on the deck until they spelt the words,

THE MARIT BUCK

"Put these on the bow and stern," he said to Dennis, "and since you are so interested in that name, you will have the job of keeping them polished until we get to Chambers Island."

On the first anniversary after John's departure
from Chambers Island, a chubby little boy with
his dog stood on the northern-most point of the
island. He had posted himself there as a lookout
early in the morning. Now in the middle of the
forenoon, he saw a black speck rise above the
horizon toward the northeast. He watched it
for more than an hour as it grew in size and took
the shape of a vessel. Finally, he saw it was a
schooner with every sail spread, churning the
water to foam with its prow as it came directly
toward the island. Then the boy ran as fast as his
short legs could carry him down the woodland
path to the house. "Ma," he shouted when he
was still far from the house. "ma, here comes pa
and the schooner." And wild with joy he rolled
over and over on the ground. Quickly Marit came
out of the house, a little girl hanging on to her
dress, another little one on her arm. At the same
time the schooner sailed into the cove and hove to
with a proud sweep so close to the shore that the
little girl jumped with fear, but Marit smiled
happily. She waved her hand and quickly was
answered by a roar from the little brass cannon in
the vessel's stern. Thereupon the cheerful rattle
of the anchor chains was heard. John stood by the
rail, looking with shining eyes upon the silent
forest island and upon the woman on the shore.
Blue were her eyes, blue was her starched dress,
and blue was the sky above her head.

CHAPTER X

EPHRAIM: A VENTURE IN COMMUNISM

Perhaps in this neglected spot is laid
Some heart once pregnant with celestial fire;
Hands that the rod of empire might have swayed,
Or waked to ecstasy the living lyre.

THOMAS GRAY.

The ordinary reader will find very little of the epic quality in the history of the average American village. As a general thing it represents only a materialistic impulse guided by chance. A hopeful grocer puts up his booth at the corner of some crossroad. A blacksmith plants his anvil on another corner, and in due time they are joined by the butcher, the baker, the candlestick maker or others of their kind, until the village is a fact. There is very little of historical import in such a chronicle.

The reader may, therefore, wonder what there can be worth telling in the history of this little northern village of Ephraim. Will not that, too, be a mere recital of drab and disconnected incidents?

But the story of how Ephraim came to be, is different. It is a narrative of unusual romance, of idealistic vision, and of sweet piety. In the evolution of its stern struggles it is kept in line by a historical unity and animated by a high purpose.

136

In the southeastern corner of Norway, near the little city of Frederikshald, lies the estate of Röd, the patrimony of the Tank family for many generations. Its far-reaching fertile fields, tilled by scores of industrious tenants, bear witness of its wealth. Its spacious parks and ancient game preserves tell of its pleasures, and its dignified manor house, filled with treasures garnered for centuries, speak of its luxuries. In its spacious halls Royalty has often found a comfortable resting place, and many cabals of state have here been constructed and unraveled.

A hundred years ago the owner of this estate was Carsten Tank, a descendant of a long line of distinguished noblemen. He was at the time the wealthiest man in Norway. He owned more than a hundred farms and estates. He had mills and factories and vast forests, while his fleet of vessels carried his produce to many foreign countries. He was also prominent in politics and for a time held a position equivalent to Prime Minister (*chef for det Förste Statsraad*).

This man had an only son, Nils Otto, who was born in the year 1800. He was a gifted young man who was given the very best education, supplemented by leisurely journeys to foreign universities and centers of culture. Nothing was too good for this handsome son of one of Norway's most powerful nobles. It was the chief aim of the ambitious old statesman that his son might be a worthy successor of his distinguished ancestors who, for centuries, had carried the name of Tank

with dignity. It is even told that he had thoughts
of placing his son on the throne of Norway. After
Norway had separated from Denmark, Carsten
Tank had taken a prominent part in placing Prince
Christian Frederik upon the throne in spite of the
protests of the Great Powers. Later, when the
weakness of this new monarch became evident,
he had taken just as resolute a part in deposing
him. Then followed the union with Sweden, with
Bernadotte (a French soldier of fortune) as King.
This union was very unpopular in Norway, and
there was a general wish for a separation, after
which Norway would choose her own King. In this
event why should not the gifted son of this prominent politician be favorably considered? He was
the scion of proud lords, had a royal bearing,
possessed learning and ability, and his father had
unbounded resources in wealth and influence. In
those troublous times, with their schemes and
cabals, when Kings were easily deposed and
commonwealths were traded like horses, this was
no impossible ambition.

But the hope of founding a new dynasty in
Norway failed to be realized. In the difficult
years between 1825-30 the business house of Tank
got into financial difficulties which strained the
resources of Carsten Tank. The Swedish King,
Carl Johan (Bernadotte), also proved to be a
strong monarch who won the respect if not the love
of his Norse subjects. These were bad enough
handicaps for political intrigues, but a worse one
was to develop. Word come that Nils Otto Tank,

the expected standard-bearer, had become a
dissenter from the state faith and a pietist.
The young Tank was, at the time, on one of
his frequent journeys abroad. In traveling through
the mountains of Saxony he met with a physical
injury and was given lodgment with some very
religious people in the village of Herrnhut. His
host was the pastor of these people who belonged
to the pietistic sect known as Moravians or *Unitas
Fratrum*. During his stay young Tank was converted, gave up all worldly ambitions, and decided,
like the first apostles of Christianity, to preach
the gospel to those who sat in darkness.

After several years as a student in Moravian
institutions, Tank became one of their most
capable teachers. In 1838 he was married to
Mariane Früauf, another prominent teacher, and
the daughter of the clergyman in whose house
he had been converted. In 1842 he went with her
as a missionary to the slaves employed upon the
Dutch plantations in Surinam, South America.
Henceforth for many years we see Otto Tank,
brought up amid the *bon-mots* of brilliant *salons*,
humbly and patiently teaching the gospel to the
degraded negroes of the distant tropics. His wife
soon fell a victim to the enervating climate, but
Tank labored on with great energy and ability.
He was six feet four inches in height, with a body
and constitution to match, and seems to have
enjoyed the physical difficulties of his surroundings.

Under Tank's efficient management this missionary station grew rapidly until a few years later

it counted twenty-nine preachers and teachers besides about a hundred assistants and servants. The congregation of negroes in Paramaribo, where the station was located, counted four thousand members and Tank in his frequent journeys into the interior had established a number of new centers of religious work. He also showed a superior business ability. Instead of buying his supplies in small quantities, as had previously been done when a vessel reached the harbor, he often bought its entire cargo, especially of flour and provisions. In order to make most profitable use of these large supplies he established a bakery and a general store, the earnings of which were enough to pay all the housekeeping expenses of the mission. He also built workshops, and bought land in order to build an agricultural school where the liberated slaves and the Indians of the interior could learn modern methods of farming. The vast forests of choice timber that covered the interior also kindled his enthusiasm and he laid big plans for systematic lumbering.

But his associates in the mission hesitated to support him in his far-reaching plans. The mission had originally been planned to be made partly self-supporting by making clothing for the natives, and most of the missionaries were former tailors and seamstresses. To make a pair of trousers or a shirt for a negro was a conception within their grasp, but with the timidity which is traditionally associated with their trade, they shrank from the thought of felling forests, building saw mills, factories and schools, and turning the wilderness into a civilized community. For them it was of greater importance to devote long hours to prayers and mournful reflection on human frailties than to take an active interest in the material welfare of the natives.

During his student years, Tank had taken much interest in the study of mineralogy. Now it happened that, in one of his journeys into the interior, he discovered one or more of the goldfields which, many years later, made Surinam an El Dorado for fortune hunters. Tank said nothing about his discovery at the time, believing that a gold craze would completely demoralize the work of the mission.

During these years Tank had more and more come to realize that the worst obstacle to the spread of the gospel was the brutal treatment which the wretched slaves suffered. He endeavored to make this plain to the plantation managers and appealed to their humanity; but these callous slave drivers would not listen to his suggestions. Accordingly in 1848 he determined to sail to Holland and try to enlist sympathy and help in carrying out his views. He made an earnest plea both to the King and his ministers on behalf of the slaves. These men were all favorably disposed toward his suggestions, and promised immediate reforms if the complaints were found to be true.

This matter caused great trouble in Surinam. When the plantation managers learned of Tank's representations to the government they became

very bitter since the suggested reforms would
greatly lessen their power over the slaves. They
therefore went to the managers of the mission and
told them that unless they defeated the purposes
of Tank they would be denied all further opportunity for missionary work among the slaves.
The managers in this way got into an extremely
disagreeable predicament. They solved it by
sending a long account to the Dutch government.
While none of Tank's representations were directly
denied, the whole problem became so wrapped up
in petty and pious nothings that the question of
reform was lost sight of completely. The sufferings
of the slaves continued and the cause of liberty
was retarded for many years.

Afterwards the mission managers expended
much pious verbiage upon the governing board of
the church and its papers, the purpose of which
was to justify themselves and throw all blame on
Tank.

The deplorable attitude which his former col-
leagues took in this matter made it impossible for
Tank to resume his post as superintendent. After
this he took no active part in the work of the
Moravian Church.

During his stay in Holland Tank was a frequent
guest of a very distinguished clergyman and
scholar, the Rev. J. R. Van der Meulin of Amster-
dam. He was a descendant of a long line of
prosperous art collectors and bibliophiles, and
his house was filled with a wonderful collection
of antique furniture of most artistic workmanship,

choice plate and paintings, rare bric-a-brac, and thousands of ancient books and manuscripts of great value. Great wealth had also come to him through his wife, formerly chief lady in waiting at the Court of Holland, and daughter of the famous general, Baron von Botzelaar, who had repulsed Napoleon at Willemstadt in 1797. For this service the baron was munificently rewarded by the Crown.

In the daughter of this distinguished family Tank found a congenial companion, and she became his wife in 1849, shortly after her father's death. In this way all these treasures became a part of the Tank possessions.

When Tank visited Norway upon his wedding journey, he was shown a letter which completely changed his future activity. It was written by some poor Norwegian emigrants of the Moravian faith, far out in the American wilderness, and asked for help.

The great migration of Norwegians to America was then beginning, and a few had found their way to Milwaukee, a little town on the frontier. There was no Lutheran church or congregation in Milwaukee, but some of the people of a pious disposition used to gather together for religious worship. Their leader was a man who, in Norway, had become devoted to the Moravian faith, and through his influence this group also decided to unite with the Moravians. They therefore wrote to the Moravian leaders of Norway asking them to send a pastor. As the times were hard they

also expressed a desire that the way might open
for them to leave the rough town of Milwaukee
and become farmers.

Both petitions were answered. A theological
student by the name of A. M. Iverson was
persuaded to emigrate to Milwaukee and become
their pastor. Shortly after this the letter was
shown to Tank, and he saw in it the finger of
Providence. Together with his wife and daughter
he sailed for America and reached Milwaukee in
the spring of 1850. The newspapers of the city
reported that Tank had with him a million and a
half in gold.

When Tank found that the conditions were as
stated in the letter, he set out at once to find a
suitable place for the colony to settle. For six
weeks he traveled over a large part of Wisconsin,
and finally selected a fertile stretch of land on the
west bank of Fox River. Here he made a pre-
liminary purchase of 969 acres of land, most of
which is now included in the city of Green Bay.
He made a further purchase of nine thousand
acres lying a little farther south on the same side
of the river. Thereupon he invited the entire
congregation in Milwaukee to come there and
settle, and promised free lands to all. It was his
plan, he explained, to build a communistic colony
on the pattern of the one at Herrnhut in Germany,
which Count Zinzendorf, the first great statesman
of the Moravians, had organized.

His offer was received with great joy by his
countrymen in Milwaukee, and in August, 1850,

the whole colony consisting of forty-two grown persons besides children, including the pastor, Rev. A. M. Iverson, moved to Green Bay. A Rev. Fett, a German minister of the same faith, also accompanied them.

Tank's first work was to lay out a number of building lots on both sides of what is now State Street, near the Green Bay Junction railroad station. Surrounding these, ten acre lots were laid out for farming purposes. These tracts of land were then apportioned among the colonists by lot, according to Moravian custom. A park covering about two acres was also laid out on the bank of the river. This was to be the site for the church. In the meantime, the north room of Tank's cottage was consecrated as a place of worship. The congregation was formally organized, which, together with the village, received the name of *Ephraim*, that is, *the very fruitful*.

There was a large two-story building in the vicinity erected by eastern Episcopalians as a mission for the Indians. This building was vacant at the time. Tank fitted it out with the necessary furniture, and here for six months the entire congregation dwelt in comfort and fellowship. The housekeeping was managed on the communistic plan. At five o'clock in the morning the matin bell aroused all. At half past five another bell called them to prayers. After breakfast the men separated and went to work at the various occupations which Tank had found for them; some to clear land, others to build houses and shops, while others sought the lake to fish. Being a man of education, Tank also made immediate arrangements for a school. One room was fitted out for educational purposes, and here five of the young

NILS OTTO TANK

men were enrolled as a students' class. Tank taught history, languages and science, while Mr. Iverson taught religion. This was the first Norwegian academy in America. It was Tank's plan to expand this school into a college where his immigrant countrymen could study medicine, law, theology, and science, irrespective of creeds, and thus become fitted to take an active part in building up the new land of their adoption.

I have spoken with old men who followed Mr. Tank from Milwaukee to their new home in the wilderness. They have told me of their joy in their new found rural liberty, of the ardor which animated them as they entered upon their work of building up their homes, and of the great hopes they had in the future of their communistic colony. It was, they said, a continual song of rejoicing, with each new day a stanza of bliss.

The founder, too, entered into his communistic plans with enthusiasm. He meditated on them as he wandered through the serene silence of the woods, and pondered their ultimate fulfillment as he sat in his cottage on the banks of the placid Fox River. He thought of his extensive travels in many lands, of his father's royal dreams, of his long service as missionary in tropic Surinam, and felt that here in the primeval wilderness of a new continent the Lord had shown him his true field of work.

Perhaps, he thought, he was to be permitted in some slight measure to emulate the shining example of that man of God, Count Zinzendorf, who had founded the religious community he supported, and whose influence had gone to the uttermost parts of the earth. His countrymen were every year coming to America by the thousands, destitute and friendless: he would help them from the bounty with which the Lord had blessed him.

There was no established church to minister to their spiritual wants: In his community they should find a well-ordered service and sanctuary. Their young people needed education and religious training: In his schools they should be amply provided for.

In imagination he saw the timbered solitudes give way to well-tilled sunny fields. He saw thrifty villages, merry with the laughter of romping children, and busy factories filled with contented working men. He heard the full-toned hymns of praise from crowded churches, and saw devout young men in his bible school studying the word of God, preparatory to a missionary life. And as plan and prospect opened before him, it seemed to him greater to be the steward of God for the relief and help of the needy in a far-away land, than to be the envied and uneasy head of a petty temporal principality.

A year passed in brotherly coöperation and the colony grew in numbers and resources. But then misunderstandings which had been brewing for some time began to be manifested. The German, Rev. Fett, was a taciturn, suspicious person, who had been sceptical about Tank's plans from the beginning and frequently voiced his suspicions to his colleague, Mr. Iverson. The latter was an impulsive man of great piety, but because of wide temperamental differences it was difficult for him to understand the broad character and lofty aims of Tank. It was probably also irritating to Iverson to occupy a secondary place in the colony. Under the circumstances it was easy for Mr. Fett to drop seeds of discord into his restless mind. Fett had told him that Tank had greatly embarrassed the missionaries in Surinam, almost wrecking their work there. What was the real motive that made this great millionaire bury himself in this wilderness? Was it a gigantic speculation, or something worse?

These and other insinuations from Fett were communicated by Iverson in confidential conversations with the members of the colony, and Tank noticed that they began to regard him with apprehensive doubt.

Finally Iverson, feeling his responsibility as pastor, decided to put Tank to a crucial test. He went to him and demanded that the colonists be given deeds to the lots which they were occupying. Tank replied that the communistic plan of the colony made this impossible. He had not left the busy activities of Europe to become a land agent in the wilderness, but to carry through a great plan of brotherly love and coöperation.

This was enough for the impulsive, inexperienced Iverson. He went to his flock and said: "He intends to make tenants of you!" No more was needed. The Norwegian tenant system was just as obnoxious to them as slavery. To be sure, they had an uncertain impression that they lived in a free country; but what can one not do with money? It was best to get out of such a doubtful situation without delay. To most of them the whole experience had seemed like a dream. That a strange man from the upper class of European aristocracy with fabulous wealth at his disposal should suddenly appear among them and give them land, church, school, and other things, seemed too good to be true. Somewhere in the scheme, they thought, there must be some snare. The result was that practically the entire colony decided to remove. Not until almost fifty years later did Iverson begin to realize the great idea which he in his impatience had assisted in frustrating. A great man had appeared among them, but they knew him not.[1]

The colony, having lost its cohesive character, was now in danger of disintegrating. Some emigrated to another wilderness farther west. Others went to Sturgeon Bay and became the first settlers there. Iverson also went there on a tour of inspection, but did not think that the soil had good agricultural possibilities. Another member by the name of Ole Larson had gone to Eagle Island, seventy-five miles north from Green Bay, and he told Iverson of the excellent fishing and of the fine hardwood timber on the mainland opposite Eagle Island.

About the same time Iverson received a loan of five hundred dollars from Bishop H. A. Schultz in

[1] Almost fifty years later the author was sitting in Mr. Iverson's house listening to his account of Tank's communistic colony. He was biased in his viewpoint, but evidently desired to be fair. When he finished his story he sat for a while in silence, then added with sadness: "I suppose I was much to blame. I was young and did not understand him. How different things might have been if we had not been so blind!"

Bethlehem, Pennsylvania, to be used in buying land for the colonists. Greatly cheered by the receipt of this money and by the news brought by Larson, he determined to go and see the land that Larson had described. About the first day of February, 1853, when the ice was safe for traveling, he set out afoot accompanied by three companions. After a march of three days they arrived at Eagle Island. Iverson describes his first impressions of the vicinity of the present village of Ephraim as follows:

The next morning we felt a little stiff after our long walk on the ice, but soon I was out of Larson's house and gazed to the southeast toward the land at the head of the deep bay. Soon I discovered that although the trees along the shore were evergreens, the timber behind was hardwood and quite different from the timber at Sturgeon Bay. With delight I looked for some time and ruminated. Perhaps our loved little congregation should be planted here on this land by the romantic bay and with the high cliff opposite so grand in appearance. After morning worship and a good breakfast, we set out with Larson in the lead over the smooth ice across the beautiful bay, a distance of about two miles. I had in my morning prayer prayed to the Lord earnestly that our investigation might be crowned with success and that we might find a good place for our congregation. With this hope I now hastened forward, well supplied with maps and diagrams. I set my foot upon the land for the first time in the name of Jesus, silently, but with strange feelings. We found after we had come to land, right close to where I later built my own house, that there was a belt of mostly small evergreens, mixed with deciduous trees, along the shore. But this belt was not broad. Soon we came to a beautiful stretch of timber, mostly hardwood (maple, beech, ironwood, with some basswood and oak), and the farther east we went, the more beautiful was the forest, the trees so high and straight and so open it was

between them that it seemed to us that without clearing a
road one might drive through it with horses and wagon
without hindrance.   About half a mile back of the shore
Brother Jacobs removed the snow, which was about a foot
deep, and dug into the soil with a stick of wood. He brought
his hands up full of black soil which he said was not only good
but rich for farming. Larson praised it no less. Strangely
enough Brother Jacobs did not this time strike any stone and
least of all did we dream of that layer of limestone which lay
only a few feet under the surface.

When we reached the shore Larson pointed out to us that
we ought to find marks upon the trees showing where the
section line ran. After some searching we found this down by
the shore close to where later our church was built. While the
brethren made this mark more conspicuous to the eye, I
stepped aside among the small evergreens and kneeled down
upon the white snow. The Saviour only knows how deeply
I prayed for the first time upon the spot. I received assurance
that right here would our Lord plant His congregation and
never forsake it in spite of all humble circumstances. I returned to my companions and told them I was solemnly
assured that here was the place for our congregation and to
this they fervently agreed. Well satisfied, we in the afternoon
returned to Larson's house, where a good dinner awaited us.

Iverson now made a trip to the United States
land office at Menasha, where he bought 425 acres
at a total cost of $478. He then platted the tract
into village lots about an acre and a half in size
with larger farm lots in the rear. In the apportion-
ment of the building sites as well as in the plotting
of the village, Iverson followed Tank's procedure
at Green Bay. The congregation also adopted the
name Ephraim, which Tank had chosen for the
new village.

One day in May, 1853, a vessel tied up in front
of the Tank Cottage to take the colonists off to
their future home. The day was radiant with the

EPHRAIM

promise of spring, but it was the darkest day in
Tank's life. Down to the vessel he saw the deluded
emigrants hurry with their few earthly possessions.
Their children carried their simple home-made
tools, their poor wives struggled with the heavy
emigrant chests, and the men shouldered their
sacks of potatoes and grain, and brought their few
cows and chickens on board. As Tank looked on
their honest faces, pinched with poverty, and saw
the heavy movements of their limbs, stiffened by
excessive labor, now about to carry them off to
greater privations and toil, they looked to him like
wayward children, sulkily denying themselves a
gentle father's care. And yet how his heart yearned
for these people! How gladly he would have
gathered them in his arms like a hen gathering her
chickens under her wings, but they would not!

But he could not follow these people. They had
spurned his gifts, and to urge further kindness
upon them would but confirm them in their
suspicions. Their paths and his had no future
crossing. Nor would he return and take possession
of the ancestral hall in Norway. His complacent
relatives smugly intrenched in pharisaic conven-
tionalism, who with pity had seen him give up the
honors and pleasures of a brilliant career to become a missionary to the slaves of South America,
would see little additional honor for him in being
jilted by a lot of praying emigrants. Better a
secluded life on the banks of the Fox, where there
was time to ruminate on the futilities of life.

And there Tank remained until his death, with

the exception of a few trips abroad for the education of his daughter.

Disappointed in philanthropy, Tank now turned to business, chief of which was his share in building the Fox River canal. In those days, before the railroad had become a recognized success, water transportation was the great problem and canal routes were everywhere surveyed, chief in importance of which was the Fox River-Portage route, the old highway of the Indians and *voyageurs*. Millions were spent on this enterprise in the expectation of being reimbursed by state lands, but the Legislature refused to recognize the claims of the company, and Tank, with others, suffered very heavy losses.

In the midst of the protracted annoyances incident to the settlement of the canal affairs, Mr. Tank took a sudden illness and died in 1864.

Very few men knew Mr. Tank. His antecedents, scholastic training and experiences of life, all made him averse to confidential intercourse. On the other hand, his old neighbors have not yet got over their awe at his patrician bearing and perfect presence, which debarred them from treating him as an equal. With his studious mind and splendid library he found more pleasure by his fireside than in the outside world. At the time of his death, he had written extensive memoirs, throwing light on the political game at the Norwegian court of his youth, as well as explaining his connection with the Moravian colony of Green Bay. He also had essays on the topography and minerals of Surinam of much importance, with reference to the gold beds of that country. These writings were subsequently to be published, but his wife, harassed by business cares, deferred the matter. Later she became so dispirited by being frequently victimized by bogus claims of charity and confidence games, that, fearing unfavorable publicity, she caused all her husband's letters and writings to be destroyed.

By many Mr. Tank's life was looked upon as a failure, and, considered as a tragedy of miscarried hopes, it was. As were the ambitions of his father, the stern premier, so were the endeavors of his son, the scholastic pietist. But his failures were more pregnant with the elements of progress than are the successes of most men.

His daughter and only child, a gifted young woman, died in 1872. His wife passed away in 1891. Upon her death the will provided that one hundred thousand dollars be given to the Tank Home for orphan children in Oberlin, Ohio. The balance, a smaller amount, was distributed to different missions. The furnishings of the house were of such rare excellence that an art expert from New York was sent for to manage their sale. He shipped the more valuable paintings, rugs and furniture to Chicago where they were sold at auction, attended by art dealers from all over the country. A vast number of smaller articles including porcelain, bric-a-brac, linen, old country copper utensils, etc., were sold at auction in Green Bay for trivial amounts. Thus the splendid collection of centuries was scattered everywhere. So

much of the famous Wedgwood ware was acquired
by the women of Green Bay at this auction that
the city is said to have more of this porcelain than
any other city in America. Here also in the public
library is one of the chief pieces of furniture from
the Tank home, a magnificent cabinet of unsurpassed workmanship, which eminent *connoisseurs*
pronounce the finest example of marquetry work
in America.

The Tank library, numbering some five thousand volumes, had been presented to the Wisconsin
State Historical Society in 1868. It was information gleaned from these books which settled the
boundary dispute between England and Venezuela
during President Cleveland's administration and
thus averted a threatening war. The Tank
cottage has been purchased by the city of Green
Bay and is now used as a public museum in one of
the city parks.

An attempt has been made above to give a brief
glimpse of this strange man Tank, who is so little
known here and by historians in Norway not at
all. For this reason his personality is almost a
myth. We are most of us "like ships that pass in
the night and briefly speak each other in passing,"
and Mr. Tank especially was not among those
who wear their hearts upon their sleeves. There
were no hurrahs and hail fellowships in Mr. Tank's
life, but rightly understood it is an epic of large
events and noble aspirations whose memory would
add dignified lustre to the historical traditions of
any community.

CHAPTER XI

EPHRAIM: A STRUGGLE WITH THE WILDERNESS

When Manitou was young and strong,
(So ancient legends do us tell),
He set about to make a home,
Where all good Indians could roam,
And peacefully in pleasure dwell.

He searched the shores of Michigan
For every pleasant cove and glen,
For towering cliffs and headlands bold,
For islands fair as toys of gold,
To make a paradise for men.

He brought these treasures to a place,
(Door County is its present name),
And here he worked with skill and might
To make a land of keen delight,
And stocked it well with fish and game.

When all was done he marked one spot,
Immaculate it seemed to him,
Where curving shore met limpid sea
In one full sweep of harmony.
What place was this? Ah! Ephraim!

The finest natural harbor on all Lake Michigan
is Eagle Island, just outside of the present village
of Ephraim. On the north the island is exposed
to all the storms that blow, but on the south,
opening toward the nearby mainland, is a cove of
deep water shaped like a horseshoe. For this
reason the island is also called Horseshoe Island.

158

It is the very exposed position of Eagle Island which has made it such a safe refuge for the sailor. It is a limestone rock, about a quarter mile in length and width, rising about thirty feet above the lake level, and covered with a thick growth of timber. The cedar trees which love to grow in the moist teeth of the wind each year send their roots far into the crevices of the limestone. Eventually the stone yields to the expansive force of the vegetable cells, and its fragments are precipitated upon the beach below. Here the waves begin their grinding and polishing work upon them. Finally they become smooth pebbles and boulders which are rolled by the waves around to the sheltered south side of the island where they are deposited in two projecting arms forming the horseshoe. As the boulders which make up the welcoming arms on the south side of the island were once parts of the crags on the north side, the island is slowly though imperceptibly crawling landward.

In the meantime the island has been a haven of refuge to thousands of storm-tossed mariners. In early days it was quite common to see a dozen schooners riding out every savage north-easter in its snug harbor, for there no waves could reach them. Nearly all of the early French explorers and emissaries of the Crown bear witness to its charm, and the old time fur traders, their *bateaux* laden with pelts from the uplands and the prairies of the land of the Dakotas, had here a favorite rendezvous.

It was also for a while the only stopping place which the steamers plying between Buffalo and

Chicago had along the peninsula on their occasional trips to Green Bay.

To this island it was that Iverson led his little
band of faithful after they had turned their backs
upon the green pastures which Mr. Tank had prepared for them at Green Bay, as told in the last

The Old Parsonage in Ephraim Built in 1853

chapter. Over on the mainland, about two miles
away, lay their little village of Ephraim with its
streets and building lots to which they all had their
titles. As yet, however, the "city" lay shrouded
in its forest mantle, untenanted since the day of
creation. A number of temporary shanties were
therefore put up on the island, and here the entire
colony lived from May till November, 1853.

Meanwhile the men divided their time between fishing, clearing land, and house building. Writing forty years later, Iverson describes the first day's work on the mainland as follows:

> I remember so distinctly the first morning when we began to clear land. There were eight of us who rowed over from the island. Arrived at my lot, I kneeled among the bushes and prayed earnestly to the Lord that he would bless the work and here plant and water His own congregation. When I for the first time swung my axe over my head it was with a vivid realization of the psalmist's words when he exclaims "Here, has the sparrow found a home and the swallow a nest." Soon the first tree crashed to the ground. I had two young men to assist me. We worked with rare energy and soon our perspiration flowed like tears. In the afternoon heavy columns of smoke were seen to rise from four different places, in that we sought as much as possible to burn up the brush as fast as we made it.

About the middle of November the greater number of the colonists moved across the bay to the new village of Ephraim, where by this time four houses were erected. Among these was Iverson's which is still standing in its original shape, size, and place. It is now one of the two or three oldest houses on the peninsula.

One of the colonists was an ingenious Dane by the name of H. P. Jacobs. He had built a good house of straight cedar logs in Sturgeon Bay. This he tore down, marked the logs and made into a raft which he towed to Ephraim. When he arrived after his long pull, it was the work of only a few hours for his neighbors to carry the logs up the steep hill and lay them up in their old order. There it stood for many years, the post office and council room where all important affairs of the village were first discussed. Fifty years after its erection it became the first hotel of the village. Many summer visitors will recall pleasant days spent in "Stone-wall Cottage."

That winter there was three and a half feet of snow on the level. For thirty miles north and south of the little settlement the forest stretched unbroken, inhabited only by wild beasts whose growls were often heard in the night. There was no postoffice or store within seventy-five miles, and no church within two hundred. But the colonists lived comfortably without these necessi-ties, eking out their slender provisions by hunting and fishing, and meeting regularly for divine worship in Iverson's sitting-room.

The next summer the little colony was aug-mented by a company of Norwegian immigrants directed to Eagle Island by a friend of the colony who had moved to Chicago. There were about fifteen families of them, and for a time they all lived in the shanties which the Moravians had erected on the island. Unfortunately they brought with them the germs of the dreadful Asiatic cholera, and an epidemic broke out. There was no physician on the peninsula and no remedies of any kind. One after another became sick, and many died. Seven cholera victims were laid away in the stony soil of the island without coffins or priestly rites. Eventually the survivors pre-ëmpted lands in the vicinity of Ephraim, and resolutely began the toil of carving farms out of the tangled wilderness.

These new settlers were very different from the Moravians. Strong of brawn, if not of brain, they were used to hard work and expected nothing else. They were not inclined to religious musings, and their idea of a pleasant holiday was one marked by boisterous carousals. John Thoreson, who after a while opened a store in Little Sister Bay, usually had a barrel of whiskey on tap, and all visitors were invited to help themselves. The story is told that some of these hearties had been on a visit to Ole Sorenson (in the present State Park). It was dark, they were drunk, and were prowling along the edge of Eagle Cliff trying to find a place to climb down and reach the beach. Torkel Knudson, who was very strong, suggested that they take John Anderson, who was tall and lank and use him for a sounding line. If he could reach the incline below, he figured it would be safe to slide down. This proposition was adopted with acclaim. Torkel seized John by the heels and dropped him "overboard." The first time this was done they tried it just above the big cave, and poor John, the protesting sounding line, was frightened into sobriety by finding himself dangling head downward over a perpendicular cliff a hundred and fifty feet high.

Such doings naturally shocked the pious Moravians greatly. Little by little, however, these people came under Iverson's gentle influence and many of them became very good church members.

During the first years there was sometimes great want in the colony. The nearest place where supplies could be purchased was Green Bay, seventy-five miles distant. But there were no roads, not even a path through the primeval jungle. Sometimes it was necessary for some of the men to walk that long distance, following the stony beach, and carry a sack of flour home on their backs. But it was exceedingly toilsome to walk on the beach, and it took a week to make the trip. During this time it was necessary to camp out every night, and there was danger of getting the precious flour wet.

Late in the autumn the settlers used to send a committee to Green Bay to purchase what supplies were needed for the coming winter. A small vessel was then found to carry the supplies to Ephraim. One fall the vessel that had been engaged for this journey was delayed by other trips. Day after day the pioneers watched to see it come around Eagle Point. They were waiting for their flour, their coffee, their salt, and a score of other household necessities. Their clothing was worn out and their children were in need of shoes and underwear. But no sail was to be seen. Finally Christmas came, bleak and bare, with none of the common holiday extras. A committee was then sent off on the long tramp to Green Bay, along the frozen beach, to learn what had happened to their vessel. When they finally arrived they found that the vessel had frozen in ten days before, just as it was ready to leave the harbor.

That winter many of the settlers had nothing on which to live but potatoes and fish. Fish for breakfast, fish for dinner, fish for supper. Occasionally the menu was varied when they were unable to catch the fish. And no salt. A few had cows, but most of them went dry because of lack of fodder. The ice was very rough that winter, but a couple of times some of the most hardy ones would set out for Green Bay as if on a polar expedition, cut their way through the ice drifts, burrow in the snow, and bring home a few of the most needed articles on a hand sled. Most of the settlers, however, received none of these spoils. They spent their time when not fishing in contriving strange footgear out of birch bark and moss, and in baiting clumsy traps with frozen fish, usually in vain. That winter an old gunny sack was treasured as a priceless fabric.

It is strange what expedients people will use when in need. About this time was born Cornelius Goodletson, who is still living, a hale and jovial citizen. After his birth, his mother had some ailments with her breasts, and was unable to nurse her baby. They had no cow. "Doctor" Jacobs was consulted. He could not relieve the mother, but he suggested that they mix some of the cheap black syrup, which was then in vogue under the name of "niggersweat," with water and give it to the baby. The mother was extremely doubtful whether the child would survive such an unnatural diet, but he belied her fears. One day the father came triumphantly home with a cow which he had persuaded someone to give up. But by that time the baby had become so addicted to his diet of "niggersweat" that he indignantly refused to

take the milk. He kept on growing and in time
became six feet three inches tall and the father of
a dozen children.

By this time so many people had settled in
Ephraim that Iverson's sitting-room was insufficient to accommodate all who came to attend
the regular religious services. But the people were
very poor and could not provide the money to
build a church. In the summer of 1857 a gift of
money was received from Rev. H. A. Schultz,
which he had collected in Bethlehem, Pennsylvania, to be used for a church building at
Ephraim. This was such an encouragement to the
people that, with much self-denial, they subscribed
a considerable sum, and the building of the church
was started. With their characteristic veneration
for sacred things, it was agreed that their little
temple of worship should not be built of the rough
logs of the forest, such as they had used in the
construction of their humble homes, but must be
built of sawed and planed lumber of excellent
quality, and in such a manner as would dignify the
church for religious use for generations to come.
Accordingly Captain Clow of Chambers Island
was sent for to go to Cedar River, then the
principal lumber port on Green Bay, with his little
flat-bottomed schooner, *Pocahontas*, after a cargo
of lumber. Iverson writes: "He soon came, but
was alone on board, so that on the trip I had to
serve both as deckhand and cook, as well as
supercargo, which was all very interesting." They
managed to get the church enclosed and roofed

that fall (1857) but were then obliged to drop the work for lack of funds.

The fact was that the little settlement was very near starvation that fall and winter. The crops in 1857 were a complete failure, due to excessive heat and drought, and in dismay the colonists looked forward to the winter with nothing to eat. The bank at Green Bay would not lend a dollar on their real estate. The mills of Sturgeon Bay and Cedar River were shut down on account of the hard times. They were almost without clothing and shoes. There was not an overcoat in the settlement. Their summer garments, made largely of worn out grain bags, were now in tatters. They thought of the hardships of the winter before when they were so near starvation. Now their potato bins and corn cribs and grain boxes were empty. What were they to live on?

In this dire extremity, Iverson launched a little sailboat which he had made, and started for Green Bay. When he arrived he hunted up a Mr. Gray, a good-natured Irish merchant who owned a large schooner. He told Mr. Gray of the serious plight in which the colony was, and said that if the merchant would advance the most necessary provisions and clothing, the colonists would pay for it by getting out as many cedar posts as he wished. This proposition was accepted on condition that all the men of the colony should come personally before Mr. Gray and enter into the required contract. With joy the colonists heard of this plan, and the men went to Green Bay and signed contracts with Mr. Gray. Personally, Iverson signed a contract for two thousand fence posts. He was also appointed foreman by Mr. Gray to superintend the work and guarantee the condition and delivery of the posts.

The following winter there was great slashing of cedar on the lowlands of the village. Iverson personally cut his two thousand posts and carried them on his back to the shore, where they were piled up. As there were no horses and only two oxen in the settlement, most of the others did the same. They were the choicest lot of fence posts that Mr. Gray ever received.

About this time occurred the first lawsuit in Ephraim, which was also one of the first in the county. The justice of the peace was Zacharias Morbek, a man of ready gifts but of a domineering disposition, who had things pretty much his own way in politics. A certain man north of Ephraim was brought before the justice, accused of assault and battery upon his (the defendant's) wife. The testimony developed that there was quite a mistake, in that the defendant had not been guilty of beating his wife, but his cow. This, however, made no difference to the stern judge. With ready decision he declared that both offenses were well known to the law, and that it was plain that the defendant was guilty of *cruelty to animals* which covered both specific offenses. He therefore sentenced the culprit to sixty days in jail and ordered the constable to take him to the jail at Green Bay. In the commitment furnished to the constable, the justice recited the facts as follows: "A. B. having made complaint in writing that C. D. did assault and beat his wife, and the testimony offered at the trial showing clearly that the defendant is guilty of cruelty to animals under the laws of this state: Therefore, it is the sense of this court that the defendant, C. D., be committed to the county jail for the term of sixty days, and the jailer be directed to feed said C. D. on bread and water, and may the Lord have mercy on your poor soul!"

In the spring of 1858 the colonists were over-joyed to receive a visit from their old friend and benefactor in Bethlehem, Bishop H. A. Schultz, who was accompanied by his daughter. They stayed for a couple of weeks, made the acquaintance of every settler, and so kind, sympathetic and noble were they that it seemed to the humble colonists as if they were visited by angels from heaven. Finally the day of parting came and all were moved to tears. It was Bishop Schultz's plan to go by sailboat to Fish Creek and there take the steamer to Buffalo. That morning, however, the water was a little rough, and Bishop Schultz, who had a strange fear of the water, said it was impossible for them to embark in a small boat in such weather. At that time there was no wagon road to Fish Creek, but only a wretched trail through the timber leading across two swamps always covered with a couple of feet of water. It was decided to try this trail. All went well until they reached the first swamp. Here Iverson proposed to follow an invisible little path which went north through the underbrush. Mr. Schultz, however, was persuaded that it would be impossible for them to find their way through that jungle. Finally it was decided that Iverson should carry Miss Schultz through the swamps while the bishop followed, making a desperate attempt to balance himself on a string of fallen trees that marked the path.

During the spring and summer of 1859, Iverson, as usual, made many missionary trips to Sturgeon Bay, Fort Howard, New Denmark, Chicago and other places in Illinois. Upon these trips he told of the hopes his members had of completing the church. He was able to obtain sufficient contributions so that the work was resumed. Doors, windows and seats were ordered from Green Bay, a steeple was built, and the church was thoroughly painted and plastered. Iverson personally made a massive, well constructed pulpit. Through a gift from Bethlehem, they were also able to purchase a bell. On the 18th day of December, the day set for the dedication of the first church in Door County, it stood complete, free from debt.

The 18th day of December, 1859, was probably the greatest day that Ephraim has seen. A heavy snow had fallen the night before, but nevertheless, when the new bell tolled for the first service in the trim little church, slowly moving oxen were seen to come from every direction, bringing sleighs packed with worshippers. They came, the Thorps, the Claflins, the Weborgs, and all the others from

the west; the Nortons and Jarmans from the
south; the Dorns, the Hempels, the Langohrs,
from the east; the Amundsens, the Andersons, the
Knudsons, and others from the north; and last,
but not least, the village congregation itself.
When the bell rang, the church was filled to the
last seat, a well instructed choir was in the gallery,
and the memorable service began. With more than
the usual fervor their pastor preached, and the
people, stirred partly by his ardent address and
partly by their own feelings, were moved to tears.
As they sat in their own well built house of
worship, it seemed to them such a great achievement that they could hardly believe it. They had
suffered so long in toil and tribulation, in cold and
sickness, in hunger and nakedness, that this
dedication of their own church seemed to them
to inaugurate a new era. For ten years the
congregation had been buffeted about, moving
from place to place in the wilderness, like the
children of Israel, but suffering far greater hardships than they. No manna daily fell from heaven
to feed them—they had to toil for it in the forest
primeval. When their wives or their children were
sick, there was no golden serpent hung on high
upon which they might look and be healed—they
could only pray in anguish over their afflicted ones.
Here no grand ceremonial cheered them on from
day to day with impressive pomp and the sound of
trumpets—they had to work out their own
material and spiritual salvation in solitude and
humility.

Poor, brave, self-denying, suffering, pioneer
fathers and mothers! Like the seed corn planted
in the ground perishing unseen to produce the
luxuriant life that springs from it, so these pioneers
buried themselves in the wilderness, and wore
themselves out with hard work, that their children
might have a better chance in life. But the children
of this new land, how little they appreciate the
sacrifices of their pioneer ancestors! They remember only in disdain their fathers' rags and bent
back, their mothers' wrinkles and rough hands,
and forget that these are the price of their own
prosperity.

This date marked an epoch in the history of
Ephraim. For many long years it continued in
its isolation, like an oasis in the desert, separated
from other settlements by vast stretches of un-
tracked forests, yet it prospered and grew. In
1864 the founder of the settlement was called to
another field, and was succeeded by Rev. J. J.
Groenfeldt, who did not suffer the light that had
been lit on "Mount Ephraim" to grow dim.
For almost seventy years that church bell has
tolled each Sunday morning, calling the people
from far and near to worship. For almost seventy
years a minister of the gospel has stood in its
pulpit, calling upon the people to turn their
thoughts from material to spiritual things. Such
teachings make for steadfastness of character, for
high standards of living. The dance hall and its
devotees have never found an opening in Ephraim.
No saloon has ever poured out its foul stench and

vulgar laughter upon this community. While the village and its people are not perfect, it is a clean, sweet place to dwell in, with high ideals and sterling honesty.

> *"O sweeter than the marriage-feast,*
> *'Tis sweeter far to me,*
> *To walk together to the Kirk*
> *With a goodly company!—*
>
> *To walk together to the Kirk,*
> *And all together pray,*
> *While each to his great Father bends,*
> *Old men, and babes, and loving friends*
> *And youths and maidens gay!"*

In closing this account of the early history of Ephraim, a word of appreciation for Mr. Iverson must be added. He was not only its founder; he was also its nurse and educator. He made Ephraim what it is. Like a mother watching over her child, so Iverson worked for Ephraim with unceasing diligence and love. This comes out strongly in the splendid narrative of his pastoral labors which he wrote almost forty years later, when he was an old man on the brink of the grave. Nothing can be more tender, more sympathetic, more loving, than his account of his pastorate in Ephraim. The reader gets a vivid impression of a faithful flock, living together in primitive conditions, but rich in pentecostal blessings. Often his narrative reaches sublime heights of pathos. His account of the death of his friend, Tobias Morbek, his story of the conversion of his little daughter at the Christmas festival and of her later death, are exceedingly touching.

REV. A. M. IVERSON

Besides his duties as pastor at Ephraim, Iverson also carried on extensive work as home missionary to other pioneer communities. He made several trips each year to Sturgeon Bay, Ft. Howard, Marinette, Cooperstown and many other places. Once he stayed a whole winter in Cooperstown for which he received his board and a cow. As he led his cow home on his ninety mile tramp he thought this was very generous compensation. Many of these mission stations later became strong churches able to support their own pastors.

All these journeys he was obliged to make on foot, on the ice in winter and in a small boat in summer, for there was no road or trail through the peninsula for many years. More than once, when overcome with fatigue, he staggered on over the bleak ice, wrapped in swirling snowstorms, and thought his end had come; but his conviction that he was God's messenger sustained him, and he tells of his toils and perilous adventures without self-commiseration.

After some years a trail was cut through the woods from Ephraim to Green Bay. It was not passable for any kind of vehicles except sleighs and it led over stumps and fallen logs and through swamps where the water stood knee deep the year around. Plum Bottom, south of Egg Harbor was a particularly bad swamp, a half mile wide where the water frequently reached waist high. Through these slimy morasses and over the stony, uneven forest ground Iverson plodded every month, pursued by mosquitoes. His friends in Bethlehem heard of his hardships and the young people there bought him a fine riding horse and saddle which they sent to him. Iverson was delighted with the excellent gift. Now he could sit at ease in his saddle and arrive at his destination in a fairly presentable condition. But his joy was short lived. He found that a horse required hay and oats, and this meant an expense which his slender income could not permit. In sadness he was obliged to sell his fine horse and once more he had to walk.

Finally the time came when he had to say goodby to Ephraim. He had opened three promising centers of work in Illinois and Iowa, and the governing board of the church wanted him to move thither and become their resident pastor. With a heavy heart he packed up his few possessions and said goodby to his communicants. He placed his wife and children in a rowboat, to be taken by a neighbor to a family in the present Peninsula Park, with whom they were all to stay for a couple of days. Then he went back to his house to commune for a little while in solitude with thoughts of his beloved Ephraim, whose creator and guide he had been, now to be left behind forever. He describes his emotions as follows:

Never can I express my feelings as I tarried a few hours in my home, now to be given up forever. I was quite alone with my God and Saviour in prayer. As I finally stepped out and closed the door behind me it seemed to me my heart ceased to beat, and I could scarcely draw my breath. With a bursting heart I exclaimed, "Goodby, dear home! Goodby, beloved Ephraim!" With slow, heavy steps I walked down the road

which for a half mile follows the shore, and time and again
I was impelled to turn and cast another look at the humble
little village, the most precious spot on earth to me. Just as
the road turns into the woods which screened the view of the
village, I burst into tears and in bitter sorrow called out a
last farewell. Oh, that unforgetable day, July 6, 1864!

Most fortunate, indeed, was this community in
the wilderness to have such a man as its spiritual
guide and friend. True, he was of a suspicious
nature and somewhat intolerant in his religious
views. But these blemishes were more than
counterbalanced by his meekness and by that
fervent devotion to his duty as he saw it. He spent
no time in looking toward the world with hopes
of personal aggrandizement. His work lay with
these pioneers, and to this work he was more
than reconciled. He was joyful. He shared their
physical labors with them, he untangled their
business difficulties, he watched by their bedsides,
and eased their sufferings with home-made remedies, he prayed for them and with them at all
opportunities.

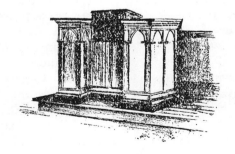

CHAPTER XII

FISH CREEK

A funny lot of folks there be
A-living in our alley,
From battle-scarred old roustabouts
To charming, sparkling Sally.
And some are crude, and some are shrewd,
And some just full of tattle.
But some are true as tempered steel,—
Fit men to fight Life's battle.

ANON.

Once upon a time there was a New York Yankee by the name of Asa Thorp. He assisted his father in tending the locks of the Erie Canal at Lockport. At that time (in the forties) a very large part of the traffic on the canal consisted of immigrant passengers bound for the West. They were packed in huge, flat bottomed scows, much the same as those now used in freighting stone on the Great Lakes. These scows were pulled along at a very slow speed by a mule walking on the bank of the canal on either side. Every day there passed one or more of these scows, loaded with stocky Germans, tall, blue-eyed Norwegians, or hopeful Irish, and piled high with all manner of painted chests, carpet bags and bundles. It seemed to young Asa, judging by their numbers and by the variety of their strange and outlandish garb, that all the world was heading for the West. Day after day they passed by, a mighty army of toilers,

178

mostly young people, determined though weary, hopeful though ragged.

What strange attractions that mighty, mysterious West had to draw so many people from the ends of the earth! He began to wonder what possibilities it had for him. Tending the locks of the canal was a job for a machine and not for a man. He began to feel the call of the wild. So, being of an adventurous disposition, one day in 1844 he stepped into one of the passing scows and joined the caravan of fortune hunters bound for the distant West.

Little by little the passengers scattered, but most of them were bound for Chicago and Milwaukee. They stayed in the scows until they reached Buffalo. Here energetic agents herded them into lake steamers on which they passed up Lakes Erie and Huron and down Lake Michigan to Milwaukee. Here in a crude little town of unpainted shanties and mud, filth and riot, they were routed out and left to their own resources.

Back in Oswego Asa Thorp had learned the trade of making butter firkins, tubs, and similar woodenware. Being desirous of seeing the country, he soon started out on a pioneer road that led into the wilderness, paying his way by making butter firkins. The road soon dwindled into a path, and after a while was nothing but a blazed trail through the timber. But along this blazed trail he would every little while come to the cabin of a new settler, and everywhere the butter-firkin man was welcome. He would stop for a few days with each settler, make up his needed stock of woodenware, inquire into the conditions of the land round about, and then push on to the next settlement.

Finally he came to the last little settlement in Dodge County called Rubicon, a few miles west of the present city of Hartford. Here the blazed trail ended and what lay beyond was a sealed book to all. However, the soil was here so fertile, the timber so tall, the conditions so promising, that Asa Thorp was well satisfied to go no further. He selected forty acres of land that suited him best. Then he hurried back to Oswego, for there was a young woman by the name of Eliza Atkinson who took the keenest interest in the outcome of Asa's journey of discovery.

Back in Oswego Thorp waxed eloquent about the wonders of the distant territory of Wisconsin. He told of the fat soil, the gently rolling land covered with huge oaks and maples, and told of his own selection of a home for Eliza and himself. The result was a rousing wedding participated in by all the members of the clans of Thorp and Atkinson. This was followed by a general exodus from Oswego of nearly all the members of the two families. To Rubicon they went and raised their cabins in the wilderness.

The land office at that time was in Menasha. Asa Thorp, being the most experienced in western ways, was delegated to go there and make formal entry of the lands. He started out and again became a maker of butter firkins. When he came to Menasha he found it was a small village on the banks of a large river flowing northward. He was told that there were many settlers on this river and that there was quite a city about thirty miles north at the head of Green Bay. Being in need of cash, Asa decided to visit these new settlements, and earn some money by his trade before returning to Rubicon. He followed the river down and met with success.

One day he was sitting in front of a store in De Pere repairing butter firkins, when a tall stranger accosted him. "Say," he said, "you ought to quit that puttering with butter firkins and come with me to Rock Island and make fish barrels. There you'll find the boys that have the cash."

"Rock Island?" said Asa, "what county is that in?"

"Dunno," said the stranger, "we ain't got no county down there."

"What state or territory is it in?"

"Dunno that," replied the stranger, "and what's more, don't care. We have no state, county or town organization, we pay no taxes, we have neither lawyers nor preachers, but we have fish and we have money. It will keep you busy twenty-four hours a day to make fish barrels at your own price. If you want to make money come along with me. It is about a hundred miles down the bay and I have my own boat."

This sounded pretty good to Asa. Big earnings and no taxes. The result was that he went with

the stranger, whose name was Oliver Perry
Graham, to Rock Island.

He found the conditions on the island as Graham
had described them. There was a large community of prosperous fishermen and they hailed
the coming of the cooper with joy. While they all
could make fish barrels at a pinch, it was beneath
their dignity when money was plentiful to handle
other tools than their fishing outfit. Asa would
have settled there for good if it had not been for
Eliza back in the woods at Rubicon.

Late in the fall of 1845 when most of the fishermen left the island to spend the earnings of the
summer in Milwaukee and Chicago, Thorp also
pulled out. He obtained a passage on one of the
Buffalo steamers that made occasional trips to
Green Bay. On the passage he got acquainted with
the captain who told him of the difficulty of
running the boats because of the lack of fuel.
Wood was used for fuel and while the entire Door
County peninsula which they were passing was
one vast forest, there was not a pier from Washington Harbor to the head of the bay where they
could take on a dry stick. Sometimes steam failed
and much time was lost sending the crew along
the beach picking up snags and driftwood whereon
to limp along until they could make port.

As the captain was telling his troubles, they
were just passing the place where the smoke from
Increase Claflin's newly built cabin could be seen
rising above the tree tops. This was the only cabin
on the entire peninsula north of Little Sturgeon

Bay and stood on the point of land opposite the cliff beneath which the village of Fish Creek was later built. "Now, there," said the captain, "is just where a man could build a pier and earn lots of money by supplying the steamers with wood."

This suggestion at once took root in Thorp's shrewd Yankee mind. He made a sketch of the indentations of the shore line, and when he reached Menasha compared his sketch with the government maps and recognized the harbor which the captain had pointed out. This done, he filed preëmption claims on all the land south and west of this harbor for a considerable distance back.

He was much elated when he reached Rubicon and told of the coup he had made. It was his intention to return to his harbor in the spring and build his pier. But hard luck and unexpected difficulties developed and he was obliged to remain in Rubicon for many years. Meanwhile his dream of riches in the vast timber resources of the peninsula floated before his vision like the thoughts of an unattainable paradise. At last, in 1853, he was able to move and build his pier, the first between Sturgeon Bay and Washington Harbor. He acquired seven hundred acres around Fish Creek and gave employment to many men cutting cordwood for the passing steamboats. Soon Fish Creek became an important business center.

Asa Thorp was a gentle-minded, capable man, always doing the right thing without ostentation. As he had lived an upright, dignified life, he looked with calmness to the end so dreaded by all. Long before his death he caused his own grave to be dug, building up its sides with slabs of slate and covering it with a slab of rock.   Here he was prepared to go

> not like the quarry-slave at night
> Scourged to his dungeon, but sustained and soothed
> By an unfaltering trust . . . .

CLAFLIN'S POINT AND CEMETERY
OPPOSITE FISH CREEK

This detail being attended to, he went about in his work of being a useful citizen in his quiet, unobtrusive way.

The Fish Creek community is not the product of a high, far planning purpose, such as is the case with Ephraim. It was the accidental meeting place of a number of discordant individuals, unrelated mentally or physically, who were driven thither by fortuitous circumstances. One thing they had in common, however, and that was the bitter struggle of finding their way through the world and battling with a merciless wilderness. With illusionary optimism they moved hither and thither, ever hoping that at the next turn they would find the pot of gold at the end of the rainbow. As an example may be mentioned Stephen Mapes. From Sheboygan he came full of hope moving all his earthly possessions, including fourteen children, for two hundred miles through the timber on a two-wheeled cart drawn by two oxen. When he came to Sturgeon Bay there was neither bridge nor ferry. But he made a raft and managed to get the oxen and all on board. With his wife standing in front of the oxen feeding them corn to keep them quiet so that none of the fourteen children would be spilled out, he paddled them across to the promised land where riches and happiness were soon expected, but which, alas, were never realized.

In time this heterogeneous gathering learned to pull together enough to be able to establish a school and even a church. This last was the work principally of two estimable women, Mrs. Griswold and Mrs. Jeffcutt, of Episcopalian persuasion. With the help of friends from the East they purchased the unfinished dwelling of a fisherman and had it remodeled. They built better than they knew, for this little chapel has a suggestion of peaceful sanctity about it which many a costly temple has failed to acquire. For a time a resident rector held regular services there. The Episcopalian form of worship seemed, however, to lack

that element of dogmatism which a hard-fisted
pioneer community seems to crave.   This was
found a few years later when a zealous Seventh Day
Adventist arrived and held stirring revival meet-
ings, centering on the saving grace of Saturday
as Sabbath. His labors were amply rewarded and
in the early spring of 1876 thirty-four grown
persons were baptized by immersion amid the
bobbing ice cakes.

In time Fish Creek achieved the distinction of
having four churches, none of which have a pastor
and the congregations of which are only dimly
visible.

Fish Creek has always been a well behaved
village, and it is many years since any saloon has
been permitted in the village or town. In the early
days a saloon was in operation where the villagers
would meet to swap fish stories over a glass of
stale beer. This public forum came to an abrupt
and dramatic end through the energy of a resolute
lady of the village, the forerunner of the famous
Carrie Nation.   One Sunday evening as some of
the village notables were dozing over a quiet game
of penny-ante, the door suddenly flew open revealing a woman with a basket full of cobblestones.
She wasted no time in words but let fly a cobble-
stone at the barkeeper. Being a woman she missed
her mark but struck and shattered the smoky
lamp.   Thereafter darkness and pandemonium
ruled the room. The lords of the card table forgot
their dignity and dived head first under the billiard
table while stones and curses flew through the air.

A door finally opened to the barkeeper's kitchen, when, seeing this avenue of escape, the men stood not upon the order of their going, but flung themselves out all in a heap, leaving the doughty woman a defiant victor.

This virtuous woman whom we may call Mrs. Squeak, had a husband. To all observers he seemed a meek and estimable man and his memory is respected. But to Mrs. Squeak he seemed to be filled with wickedness, and she was everlastingly on the warpath to drive it out of him. The chastisements that man received are the wonder of all old settlers. Once he and a few others were having an innocent celebration in a fish shanty. Someone was caressing a fiddle, another was demonstrating what he did not know about a jig, and Mr. Squeak, utterly forgetting his domestic experience, was singing with a woeful voice, "Oh, I'm a ti—ger! Oh, I'm a ti—ger!" It was a scene of blissful contentment.

Then the door opened, revealing Mrs. Squeak with a whip in her hand. For an instant she viewed the paralyzed inmates with a baleful glare. Then she squeaked: "Well, if you are a tiger I'll put the stripes on you!" Whereupon she unmercifully lashed the cowed tiger to his cage.

This woman was a terror to all sinners. One withering look from her eyes was enough to blast the sunshine out of any ordinary man, dry up his laughter, and shrivel him into a speechless nonentity dressed in rags. And when she opened the floodgates of her wrath the stoutest hearts in the

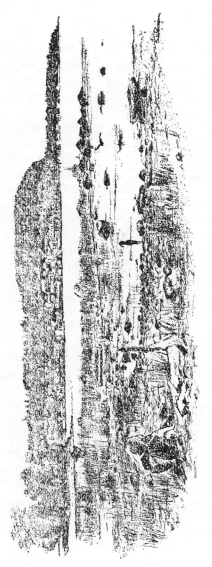

FISH CREEK

village forgot dignity and forthwith sought refuge in flight. Once, however, she met a disastrous defeat.

There was a good natured elderly woman in town who, for a time, was dock agent. It was therefore necessary for Mr. Squeak, who was a business man, to have some conversation with her. Mrs. Squeak felt convinced that these conversations were filled with unspeakable evil and she longed to annihilate this smiling temptress of her spouse. One Sabbath morning, attired in all her finery, she encountered this woman's husband. Being a lazy man he had just milked and was crossing the road with a pail of milk in his hand. She stopped him and gave him a piece, a long piece, of her mind. She told him what she thought of his own low-down self, his unspeakable wife, his worthless children, and his despicable ancestors as far back as she had ever heard of them. He received this cloudburst meekly, with a bowed head, as befits a mere man when female authority speaks. But when she stooped down in frenzied indignation, picked up a handful of mud and threw it into his milk, the worm turned. He looked into his grimy milk pail a moment and then said: "Well if you want mush and milk, you can have it right on the spot!" With that he doused the whole milk pail over her.

She gasped and choked. "You—you—you—!" she spluttered, but speech for once failed her. She turned on her heel, trailing milk, misery, and mortification to her home.

Ezra Graham was his name. Ezra Graham. It is only meet that the conscientious chronicler give all proper credit to such heroism, no matter how misguided.

There are here and there people who by their disregard of accepted standards of conduct become public characters. Gossip does not create their reputations. They have, as it were, shouted them from the housetops, and their record becomes a part of the traditions of their community.

There are also people whose quaint and usually harmless eccentricities add a smile to the recollections of their neighbors. Fish Creek had many of these.

Well remembered among them is old Myron Stevens. He had a slant for law which, coupled with his wit, made him a famous pettifogger. A good time was always expected when Myron Stevens took a case. His wit always won him the good will of the jury, and his strange lines of defense always mystified and perplexed the opposing pettifogger. Sometimes, prompted by mischief, he would read a paragraph from the statutes, modifying the phraseology and interlarding it with clauses of his own invention. On hearing this strange reading of Wisconsin law the opposing pettifogger would jump up and shout "Say! Hey there! Let me see that! Where do you find that paragraph?" But old Myron would solemnly snap the book shut and haughtily reply: "I am here to defend my client; not to teach law to greenhorns."

When old Myron had anything he was very generous and when he did not, he expected others to be so. One day he came into a neighbor's house and said: "Say, John, can you let me borrow a piece of bacon? I'll bring it back when I have cooked my beans." Another time in winter he was riding along behind a slow horse, blue with cold, his teeth visibly chattering. A passerby called to him, "Say, Myron, why don't you get out and walk and get warm?" "N-no," replied Stevens in frozen dignity, "I would rather sit and freeze like a man than run behind like a dog."

The boss of the town for a long time was Alexander Noble, who had a disposition as crusty as that of a traditional English lord. He was an expert blacksmith and no one knew it as well as himself. When anyone wanted Alexander Noble to do anything it was necessary to use more circumspection of speech than if he were addressing the President. Once a farmer came in and said, "Say, Mr. Noble, can you shoe my horses?" Haughtily the blacksmith turned and said: "Do you mean to insinuate that I who have been shoeing horses for thirty years can't shoe your worthless plugs?"

"I meant to say, *will* you shoe my horses?" faltered the farmer.

"Why don't you say what you mean then? Now get out of here till you learn to speak intelligently."

Among the queer characters who helped to make things lively in the village was a "Doctor" Hale. He and his wife were traveling members of the

Kickapoo Indian Remedy Co., which was a cross between a circus and a patent medicine agency. His wife had been a performer on bare-back horses and profoundly impressed the populace and shocked Mrs. Squeak by dashing about on horseback in all manner of perilous postures. No less were they impressed by "Doctor" Hale who carried in his pocket a $1,000 bill. It is still a matter of debate whether it was bogus or genuine, but it was remarkably efficacious in winning respect or securing credit.

About this time E. S. Minor, another old settler, opened his campaign for congressman. "Doctor" Hale let it be known that he had been Senator Gallinger's private secretary and had practically "made" the Senator. He offered to give Mr. Minor the accumulated wealth of his vast experience, and guaranteed his election if he were given free hand as campaign manager. Mr. Minor's friends now felt that the election entirely hinged on Hale's coöperation and beseeched Mr. Minor not to commit political suicide by refusing to engage Mr. Hale. Mr. Minor, however, stolidly refused the potent aid of the $1,000 bill and went ahead and was elected just the same. It later developed that while "Dr." Hale was from the same state as Senator Gallinger, he had had no connections with him whatever.

Fish Creek was for a long time the principal fishing center of the peninsula. Nearly every man in the village was a fisherman, and north and south of the village the shore was lined with fishermen's homes and nets.

The first fishermen in this region (after Increase Claflin) were two stalwart Norwegians named Ole Weborg and Eben Nelson who reared their homes side by side two miles north of Fish Creek in 1852. For many years the twin lights from their evening lamps served as a lighthouse to the many schooners which passed up and down the bay. Behind them, on top of the cliff still stands the house of Sven Anderson, another fisherman and a gentle-minded old bachelor who settled here because of the transcendent beauty of the scenery. He built his home on top of Sunset Cliff, also known as "Sven's Bluff," where he could enjoy one of the finest views in America. This made it necessary for him to tote his water and other supplies two hundred feet up the steep hill, but he considered this a slight inconvenience in contrast with the panoramic feast which he enjoyed up there. His house was a Mecca for all the children of the neighborhood, because he always treated them so courteously and generously. Although a bachelor he always had a petticoat hanging on the wall, "so as to make the house look a little more homelike."

To be a fisherman in those days required much resourcefulness and efficiency. In addition to their regular work of fishing they also had to build their homes, fish huts, piers and often their boats. They also made their own nets, or had them made by some of the women of the neighborhood. For weaving a gill net, six feet wide and one hundred and sixty feet long, two dollars was paid.

A beginner at this trade would work for weeks to make a net.

But the biggest task aside from the daily fishing was to make the fish barrels in which their fish were salted and shipped. In those days every fisherman was his own cooper. As several hundred barrels were needed for each season's catch it was

a job which kept them employed all through the winter.

The first step was to get the pine timber needed for the staves and headings. Two men would usually go together back in the woods with their lunches, axes and a cross-cut saw. The pines were felled and sawed up into bolts about three feet long. Then some neighbor with a yoke of oxen would be hired to haul them out. These pine bolts were then split up into thin slats and piled up, log house fashion, in long rows. Thus they would stand all summer in order to dry out.

The next step was to dress these slats down to finished staves and headings. The cooper would sit astride of a long bench facing the other end of it and by stepping on a paddle or lever the stave was held in a vise-like grip in front of him in the proper position for working. A slightly curved draw-shave was used for planing the stave until it was about a half-inch thick and gently tapering at both ends.

Another trip had to be made back in the swamps for black ash from which the hoops were made. Each barrel required eight hoops about seven feet long. These ash logs were split into sections about two by four inches in diameter and were not allowed to dry out. An ingenious method was used to split them into hoops. One end of these split ash timbers was inserted between two fixed timbers—usually a hole in the wall of the cooper shop. Then by bending down on the other end each year's growth of the timber was made to separate from the adjoining. Each timber would thus make five or six hoop stringers of just the right thickness. The timbers were left unsplit in one end for six or eight inches so that the whole bundle could be handled more easily. These *tassels* were then put to soak in a trough about eight feet long, a foot wide, and deep enough to hold a large quantity of hooping. When needed they were taken to the shaving bench, the ends lopped off and matched and trimmed for putting on the barrel.

The headings, of which there were two halves for each end of the barrel, were planed down with the draw-shave to the right thickness. The inside edges were matched and two holes about an inch or two deep were bored into the inside edge of each half heading. This was done by means of a little boring machine which was turned with the left hand while the right hand held the wood in place. Two holes were thus bored in one operation at the same distance apart. Into these holes ironwood pegs or dowels about three inches long were driven which firmly united the two halves into one. A circle was then traced upon the heading boards by means of a compass to mark the outside edge of the heading, the unnecessary wood was trimmed off with an adze, and last a circularly faced plane was used to work down to the line and to bevel the edge of the heading.

The staves, hoops and headings being ready, the next step was to assemble the barrel. The cooper's work-bench had a half circle cut out from its front edge and below this, about a foot from the floor, was a small platform. The staves were put together on this platform and in the curved cut of the work bench. They were first fitted together inside a thick, ironbound ash hoop or frame, called the truss-hoop. When the last stave was fitted in the truss-hoop was gently hammered down, until each stave acted like the key stone in an arch to keep the others in place. They were then steady while the first permanent hoop, called the love-hoop, and the two first quarter-hoops were put on.

Every coopershop had a large fireplace and this played an important part in coopering. When

A COOPER'S FIREPLACE

one end of the barrel was solid, so there was no danger of collapse, the barrel had to be "fired." For this purpose a sheet-iron cylinder about two feet long with holes in the lower end was placed upright in the fireplace and filled with shavings and small chips. The fuel was ignited and the barrel was placed over, or rather around the cylinder, with the sprawling ends down. Here it was left exposed to the heat of the redhot cylinder until the staves were well charred, often turning black inside. This was to thoroughly dry the wood so that when the last hoops were put on there would be no danger of leakage. It also caused the staves to retain their bent shape when they were drawn together. Some coopers were very expert in this firing, coolly going about other work while a bystander would think the barrel was being hopelessly burned. While the barrel was still warm and its staves quite flexible it was returned to the work bench and the sprawling stave-ends were drawn together by means of a rope around them drawn tight with the aid of a small windlass which was held firm by means of a ratchet. The last hoops were then put on. The inside ends of the barrel were finally grooved for fitting in the headings, leaving a chimb about an inch wide.

Not a single nail or rivet was used in making these water tight barrels which were subject to much hard usage.

Adjoining Fish Creek to the southwest lay a tract of land which Asa Thorp was very dissatisfied with, after he had cut the timber off. It lay between the rockbound beach and a perpendicular cliff, a nest of stones. As far as he could see there was at least a ton of stone to every shovelful of dirt. If anything was ever created in vain surely this was. He tried to dispose of it, but no farmer would take it for a gift. They shuddered when they looked at it. Finally Dr. Welcker came to Fish Creek to start a summer hotel for a select class of Germans. Here was an "easy mark." Full of rapture over the scenery, Thorp tried to unload on him. He offered the doctor a whole mile of this stony beach-land for two hundred dollars, but the doctor was dubious.

This tract of land which Thorp could not give away is now the site of the most extensive and well kept summer colony on the peninsula. In point of beauty of location and surroundings it is unsurpassed by anything on all the Great Lakes. Nowhere can be seen such dignified stone walls or beautiful rustic cedar work as here. Its tennis courts, built on the side hill, are monumental works of art. Behind them rises the castellated cliff more than two hundred feet in the air, its broad breast adorned with many a bouquet of green cedars. In front of the cottages the limpid waters of Green Bay stretch away to the far horizon, each night presenting a sunset of gorgeous and indescribable beauty.

A jolly good town is old Fish Creek,
   The best on the pike, I know;
With its back to the rock and its face to the sea,
   Where the rollicking breezes blow.
As snug as a bug in an old woolen rug,
   It lies there embowered in green:
You may go where you like, on any old pike,
   No cozier village is seen.

When old Father Claflin discovered "old Door,"
  Some four score years ago,
With Indians and black bear it was galore,
  And sturgeon—a wonderful show!
He roamed the timber and cruised the shore,
  Delighted with all he did see;
But when he saw Fish Creek he roamed no more,
  But said, "My home here shall be!"

CHAPTER XIII

EGG HARBOR

O Willie brewed a peck o' maut,
And Rob and Allan came to see;
Three blyther hearts that lee-lang night
Ye wadna found in Christendie.

BURNS.

Once upon a time a young Chippewa Indian from Washington Island was hunting with his dog on the hill overlooking Horseshoe Bay in the present town of Egg Harbor. As he was cautiously stepping forward between the tall trees skirting an open glade, he spied two bear cubs comfortably dozing on the sunny side of a big windfall. Being, like most Indians, fond of pets, he restrained and silenced his eager dog, and crept forward intending to capture the cubs alive. When near them, he laid down his gun and suddenly pounced upon them. Immediately there was much squirming and yelping; but he managed to get a good hold on the fur of the neck, and, after a brief struggle, arose with a cub in each hand. No sooner was he on his feet, however, before he heard a ferocious growl behind him. Turning instantly, he saw a huge bear advancing upright on her hind feet. His gun was on the ground some distance away, and there was no chance to escape. Dropping the cubs, he pulled out his tomahawk and knife to defend himself. But he had only time to raise his arm to throw the tomahawk when the savage beast was upon him. A blow of her paw broke his arm like a pipe stem and sent the tomahawk flying through the air. Then she seized him in a terrible embrace, and he felt his ribs breaking.

Still clutching his knife in his other hand, he was able to give his huge adversary several ugly slashes in the abdomen. This did not seem to bother her much, and he soon would have been crushed to pulp if it had not been for his valiant dog. So fierce were the attacks of his faithful ally that the bear found it necessary to turn her attention to the dog. This gave the Indian his opportunity. He jumped for his tomahawk, and resolutely advancing drove it with a sure and heavy stroke into the skull of the bear which fell dead.

By this time the little bears had disappeared in the woods and he had to give up their capture. His companions, who were not far away, set his arm and skinned the bear. In recognition of his great bravery in attacking and killing a bear while suffering from the pain and handicap of a broken arm, the Indians called him Big Bear and later he became a famous chief.

Egg Harbor seems to have been a great place for bears. Old settlers tell of their invading storehouses to steal bacon and drink milk. Once it even happened, in broad daylight, that a child was carried away by a bear. This occurred about three miles southeast of the village, at the home of a young pioneer by the name of Fred Kraucht. The father had gone to Bailey's Harbor, and the

mother had left her two year old child to the protection of a dog that was chained near the house,
while she strolled down to gossip with a neighbor
living not far away. Soon the dog was heard
savagely barking and the child screaming in terror.
The mother and the neighbors rushed to the house,
but the child was nowhere to be found. A general
alarm was sounded and all the neighbors turned
out to search the woods and fields thoroughly.
The child was never found; but some fresh bear
tracks leading into the swamp gave a grewsome
explanation of the disappearance.

The name of "Egg Harbor" is unique, probably
not being duplicated anywhere else in America.
It dates back to a picnic frolic which took place on
its lake shore exactly a hundred years ago as this
is being written. The Hon. Henry S. Baird of
Green Bay told the story of its naming as follows:

In the summer of 1825, Mr. Rolette, a prominent and
extensive fur trader, arrived at Green Bay from the Mississippi, with three or four large boats, on his annual voyage to
Mackinac, with the returns from his year's trade. Since there
was at that time no vessel at Green Bay, he kindly offered
passage on his own boat to Mr. and Mrs. Baird, then "young
folks" who resided at Green Bay and were anxious to visit
Mackinac. On a fine morning in June the fleet left the Fox
River and proceeded along the east shore of Green Bay, well
supplied with good tents, large and copious mess baskets,
well stored with provisions of all kinds, especially a large
quantity of eggs. On the second day at noon the order was
given by the "Commodore" (Mr. Rolette) to go ashore for
dinner. The boats were then abreast of Egg Harbor, until
then without a name. On board the Commodore's boat, there
were besides himself, Mr. and Mrs. Baird, and nine Canadian

boatmen, or voyageurs, as they were styled. On another of the boats were two young men, clerks in the employ of Mr. Rolette, one of whom was a Mr. Kinzie, and a like number of boatmen. It was the etiquette on those voyages, where several boats were in company, that the principal person or owner took the lead. Sometimes, however, a good natured strife would arise between the several crews, when etiquette was lost sight of in the endeavor to outstrip each other and arrive first at the land. . . . . At the entrance to the harbor the boat in charge of Mr. Kinzie came alongside the Commodore, with the evident intention of running ahead of him. Mr. Rolette ordered it back; but, instead of obeying, the crew of the boat, urged on by Mr. Kinzie, redoubled their efforts to pass the Commodore, and, as a kind of bravado, the clerks held up an old broom. The Commodore and his companions could not brook this. The mess baskets were opened and a brisk discharge, not of balls, but of eggs, was made upon the offenders. The attack was soon returned in kind. It became necessary to protect the only lady on board from injury, which was accomplished by covering her with a tarpaulin. The battle kept up for some time, but at length the Commodore triumphed, and the refractory boat was obliged to fall back. Whether this was the result of superior skill of the marksmen on board the Commodore's boat, or the failure of ammunition on the other, is not now remembered.

The battle was renewed after landing. The boats and the men presented a rather unusual appearance, and the inconvenience was increased by the fact that some of the missiles used by the belligerents were not of a very agreeable odor. The fun ended in Mr. Kinzie having to wash his outer garments, and while so employed some mischievous party threw his hat and coat into the lake. All enjoyed the sport, and none more so than the merry and jovial Canadian boatmen. The actors in the frolic long remembered the sham battle at "Egg Harbor" and it is believed that to this rude frolic may be attributed the origin of the name of this town in Door County.[1]

[1] Quoted from the *Door County Advocate* of April, 1862.

In 1853 the first permanent pioneer came to Egg Harbor and settled on practically the same spot where the battle with the eggs was fought, that is, on "The Point" west of Alpine Lodge. His name was Milton E. Lyman, and he became a famous man in Door County. Mr. Lyman was a descendant of a prominent eastern family and no one knows how it happened that such a man of education and intelligence should seek a home so far away in the wilderness. Probably some romantic tale or some intrigue lies hidden here, never to be revealed. Mr. Lyman was a popular and companionable man, admired and dreaded for his wit and sarcasm. He became the first County Judge and was also at the same time Clerk of Court and County Superintendent of Schools. After this he was for many years Justice of the Peace and as such united no less than seventy-three couples in marriage. He was assisted by a small following of constables and pettifoggers who were great in drumming up business, always on hand to offer their services the moment a row broke out. Down to his little house on the shore a well beaten track was worn, on which was seen almost any day, a procession of pettifoggers, plaintiffs, defendants, witnesses, constables and spectators. Heated trials were held, and the "Judge" would gravely announce his decision. The amount of the fine was so nicely adjusted to the size of the crowd that it was seldom necessary for the magistrate to pay the barkeeper anything out of his fees, though it did sometimes happen that through miscalculation the defendant was obliged to pay extra for his own drink. At the frequent weddings the judge was, of course, the principal guest of honor and then there was no troublesome barkeeper to keep track of the drinks.

When business permitted Judge Lyman was also in much demand as a pettifogger in other justice courts, and as such had many sharp encounters with the valiant champion of Fish Creek law, Myron Stevens. In those days each little community was as vain of its principal pettifogger as a present day college is of its principal football star. There was therefore much good natured scoffing between the rival clans. Once a native of Egg Harbor was arrested for stealing a pig. He had chanced upon a plump suckling asleep in the corner of a rail fence. Temptation overcame him and he slipped the little animal into a pocket of his coat. Unfortunately for him, the little captive wriggled out just as he was passing through the village. Here no pettifogging could obscure the facts in the case, and the prisoner pleaded guilty. In pronouncing sentence Judge Lyman said: "I will give you the choice of thirty days in the county jail,—or three days in Fish Creek."

The prisoner groaned in dismay at the dismal alternative; but he was loyal to the core, though otherwise a reprobate.

"Gimme the county jail," he exclaimed.

It developed later that Judge Lyman was far from being as circumspect in his own conduct as was looked for in a dispenser of justice. There were features in his private life which are unmentionable in polite society. His public delinquencies were also numerous. His last public office was that of town treasurer during which he embezzled a thousand dollars. When confronted with his crime he brazenly gave the town authorities the choice of sending him to jail and losing everything, or letting him go free and taking his note which "would soon be paid." The second alternative was accepted and the note was handed down from one set of officials to another as a doubtful asset until it was practically worn out. The signer, also in time wore out, and died as a town pauper, when the note was burned.

But the people of Egg Harbor had other things to do than to concern themselves with Judge Lyman and his witty dispensations of justice. Back from the shore lay a wilderness of swamps and stony hillsides covered with dense timber. This was to be cut down, burnt up or marketed as the case might be, after which the remaining chaos of stumps, stones and brush piles was to be turned into tillable land. This was a man's job if there ever was one! Nowhere was it more laborious to subdue the land to the needs of man than in the region around Egg Harbor. Into this jungle without roads or paths, far from the comforts of civilization, the sturdy pioneers penetrated, often carrying their cookstoves on their backs. A little shanty, six by ten feet square, was the first house, with a bunk in one end. A rough roadway was next opened to the nearest neighbors. Then, while the perennial pea soup was simmering on the

stove, the pioneer was battling with huge maples
to be made into cord wood.   For this he might
get as high as two dollars per cord, after he had
succeeded in transporting it a half dozen miles
to the nearest pier. No radios then wafted to him
the music of distant cities to break his loneliness.
There were no automobiles to carry him pleasantly
around on well built roads; there was not even
the telephone or the newspaper to tell him of the
happenings in the world outside.

One is prompted to ask, what inducements
could there be to take up such a life of toil and
bleak self-denial?   But for every emergency there
seems to be men ready to meet it.   These early
pioneers were a picked lot of brawny workers with
the lust of conquest in their hearts who laughed
at hardships and never felt better then when
struggling with some impossible task.   They had
marvelous appetites and still more marvelous
digestions and the smoke from their smouldering
brush piles at dusk was as incense to their nostrils.

Nor were they without amusements. In the long
winter evenings could be heard the crooning of a
violin accompanied perhaps by the distant howl
of a timber wolf. Whenever one of these hardy
axe-men had prospered so much as to be able to
get up a new log house or barn the event was
always celebrated with a rousing party. To these
parties the backwoodsmen would gather for miles
around, since there was to be had plenty to eat and
drink—especially to drink—for whiskey was only
a shilling or two per quart, and an uproarious good

time was enjoyed by all. The fiddler was never lacking and soon the polka, the schottische, the reel and other old-fashioned dances were in full swing. Not infrequently these parties resulted in physical demonstrations not exactly sanctioned by good form, for wood chopping energies demanded unusual relaxations. The following account of one of these parties was written at the time of happening by the able editor of the county newspaper, Mr. D. S. Crandall, and gives a vivid picture of what sometimes took place.

A farmer living a few miles from the village of Egg Harbor invited his neighbors to spend a sociable evening at his home. It is not at all likely that his hospitable offer would have been refused, even though no other attraction than a dance had been promised, for amusements are always welcome in that locality, so that the giver of a party is not obliged to send out a press-gang in search of guests, as was the case of the gentleman whose marriage is recorded in the New Testament. But having backed up his invitation with the assurance that there would be plenty of beer for women, children and other temperance people, and lashings of whiskey for those who preferred to get drunk with neatness and dispatch, it is hardly necessary to say that he had a crowded house with "standing room only" for those who arrived after seven o'clock.

It will be readily understood that under the influence of abundant grog the evening had not far advanced before there was such a tremendous sound of revelry that if there had been any police in the vicinity they would have pulled the house and brought the entertainment to an abrupt conclusion. But, there being no legal impediments to the festivities, they were conducted on such a free and easy scale as would have astounded those who lived in a more civilized community. Long before midnight the fun became boisterous and decency received the grand bounce. It was while affairs were in this interesting state that one of the men, who was possibly a little more tipsy than the rest, laid the foundation for a first class row. The whiskey he had drunk excited his affectionate instincts to such a degree that regardless of his surroundings he made advances of an indelicate character to one of the women, who immediately proclaimed the fact by a squeal that drowned all other noise in the house. The woman's son was present and when he learned of the cause of the trouble he struck out from the shoulder with such vigor and precision that the offending man took a tumble under the table, where he lay for a few minutes trying to discover how many of his teeth had been loosened.

It might be supposed that a man who had committed such an offense against the moralities would have no sympathizers, and that the verdict of the crowd would be that he should be kicked as long as kicking was good for him. This would doubtless have been the opinion of the guests if they had been sober, but, being drunk, they took a different view of the matter. Moreover, up to this time there had been no fight, and all hands had taken just enough whiskey aboard to make them itch for a scrimmage. The consequence was that within two minutes every man in the room was endeavoring to put a head on his neighbor. No one appeared to know or care what he was fighting about, the chief aim of each belligerent being to put his knuckles where they would do the most good.

It did not take the ladies long to realize that the men were conducting a riot with so much skill and energy that assistance of the fair sex was entirely unnecessary. In order, therefore, to give the combatants abundant room, and also to get themselves out of harm's way, the women bundled themselves and the children off to the room upstairs. The terrific uproar below caused several of them to go into hysterics, and when their condition became known in the lower regions some of the men went to their relief. The additional burden thus put on the chamber floor was more than it could support and the joists gave way with a crash, precipitating men, women, children and furniture upon the heads of the pugnacious gentlemen on the ground floor. For about five minutes that floor presented an appearance to which no description can do justice. Many of the ladies were standing on their heads, their limbs sticking out of the heap in every direction like the spokes of a busted cart wheel, while their striped stockings waved in the air like signals of distress at the masthead of a water-logged scow. The children screamed, the women shrieked and the men swore as in their efforts to disentangle the squirming mass of humanity they found that a woman was being pulled out of the heap in different directions. When at last order was restored everybody was surprised to find that nobody was either killed or seriously hurt. The fighting party had escaped the falling floor, and the people from above were none the worse for their tumble.

The accident had at least one good result. It brought the row to an end, and now all hands were as ready to bind up their neighbors' wounds as they had lately been to inflict them. As soon as the women recovered from their fright they began to count noses to learn whether anyone had been lost. The inventory showed that one of the children was missing, and for a short time the mother was distracted. The young kid was finally discovered in a flour barrel into which it had fallen when the floor gave way, and was restored to its mother's arms along with several pounds of "double extra" breadstuffs that had powdered the infant from head to foot.

Having almost torn the house to pieces, pounded each other for about an hour, and nearly succeeded in killing the women and children, it was mutually agreed that there had been enough fun for one night. The guests therefore collected their wraps, took one more drink all around in token that they bore no ill-will towards one another, and departed assuring their host that they had spent a most delightful evening and that his party had been the most successful affair of the season.

CHAPTER XIV

THE GIANT OF HEDGEHOG HARBOR

He was six foot of man, A 1,
Clear grit and human natur';
None couldn't quicker pitch a ton,
Nor quicker eat a creatur'.
Apologies to LOWELL.

At the extreme northern end of Door County, a wide, rock-rimmed bay opens into the peninsula from the north. On the west side it is guarded by the magnificent Door Bluff, which rises defiantly two hundred feet high above the roaring sea. On the east it is sheltered by the somewhat lower Table Bluff, top-crowned with cedar and balsam. The arc of the enclosed semi-circle of gleaming waters is indented at its apex by a third bluff, noble in its castellated terraces, but nameless in the presence of its greater brothers. The west half of the bay is called Garretts Bay; the east half is known as Gills Rock.

The name of Gills Rock is a comparatively recent innovation, so named after Elias Gill, who, in the early seventies, came here with his woodchoppers and devastated the land. On old maps and in ancient days the bay was known as Hedgehog Harbor.

Almost a hundred years ago there was on Rock Island, ten miles outside of Hedgehog Harbor, a man by the name of George Lovejoy. He had

been a sergeant in the United States army and had settled on Rock Island in 1836. Lovejoy was a famous hunter in many parts of northeastern Wisconsin, and possessed a remarkable faculty for almost anything he undertook. It was said that he almost broke up the settlement on Rock Island by the bewitching, homesick melodies of old time songs he drew from his violin. He was also a master ventriloquist. Sometimes he would go out on the ice where an Indian was fishing, and make the trout talk back to its captor in the most approved Chippewa dialect, to the poor Indian's terrorized amazement.

At an early date Lovejoy built a small vessel, said to be the first one built in Door County. One autumn, at the approach of winter, his vessel was thrown on the beach near the present Gills Rock pier. Next spring when he returned to launch her, he found her so high above the water that he had to give it up. Later he returned with a companion, a tremendously strong man by the name of Allan Bradley, and with his help the boat was launched. During the spring the porcupines had gnawed so many holes in her, however, that they had much difficulty in making her float.

When Lovejoy was ready to leave, he offered to pay Bradley for his help. But Bradley would not accept anything. "I like this Hedgehog Harbor of yours so much," he said. "It is the pleasantest place I have found in the West, so I am going to build me a home here. Since you brought me here, I will take nothing for my work." Allen Bradley then built a shanty just back of where the present pier stands, and in 1856 became the first settler in the vicinity of Hedgehog Harbor, by which name it was afterward known.

Allen Bradley later became one of the epic characters of Door County. He was a good-natured, square-dealing person, more than six feet tall, but he was so broad that he looked rather stocky than tall. He measured more than four feet around the chest, had hands as broad as shovels, and was obliged to wear moccasins because no shoes could be obtained that were big enough. The old settlers in the northern part of the county speak of "Old Bradley, the timber chap, who lived like an Indian and could cut seven cords of body maple in a day." In those days crosscut saws had not yet come into use, and the big maple trees were felled and cut into cordwood with axes only. Bradley had a home made affair as heavy as a maul, and, with his strength behind it, chips weighing a pound apiece would fly at every stroke.

Bradley did not spend much time cutting cord-wood, however. He was a leisurely fellow, hunting and fishing and tapping his maple trees. Money was not much needed except for the annual purchase of flour and knick-knacks for the family. The deer that bounded through the timber gave him abundant food and the best of clothing. Maple sugar for his flapjacks in the morning, a bear-steak for dinner, a whitefish for supper, furnished a menu that did not cost him five cents a day, and which a fastidious epicure of the metropolis could not surpass in quality. Life was easy.

Many stories are told of Allen Bradley's incredible strength. He had a twenty-four foot pound-boat, a cumbersome, flat-bottomed scow, weighing more than a thousand pounds, and he and his son were wont to pull this up on shore by taking hold one on each side. Once at Washington

ALLEN BRADLEY

Harbor, nearby, six men were vainly struggling to lift a big timber into place on the crib they were building. Bradley looked at them for a while and then, when they sat down gasping after their ineffectual struggle, lifted the log alone and placed it in position.

In Escanaba he and another big fellow named Ransom Call once took a job of stacking a cargo of salt barrels. Call was a big bulky man who had the reputation of being tremendously strong. The barrels weighed three hundred pounds apiece, and were to be stacked in rows on the pier three tiers high. It was slow work to roll the barrels along and then lift them up into place. At noon, therefore, the captain, knowing what bears he had for longshoremen, promised a reward of five dollars to the one who could keep longest at handling the the barrels singly.

"Give the money to Ransom," said Bradley. "He is bigger than me."

"Oh, no, let's try for it first," said Call, confident of his strength.

They started in, each man grasping his barrel by the chimb and lifting it up to its place. When it came to raising the barrels to the third tier, it was necessary to lift them breast high, and soon Call was perspiring profusely. After a while he called Bradley aside and said:

"Say, Al, let's quit this rushing. The man who loses don't get nothing for his trouble. Now if you'll quit first, I'll give you the five."

"Quit yourself," answered Bradley, "I'm not even warmed up yet." With that he nonchalantly resumed heaving the barrels into place.

Call desperately kept up for a half hour; but as Bradley showed no inclination to stop, he finally dropped down behind the barrels, all exhausted, He was too fat. Bradley, however, kept on till the cargo was duly piled on shore.

A schooner was once wrecked on Hanover Shoal. Allen Bradley took the job of salvaging the rigging. He cleaned her up alone from keel to truck, including a thousand-pound anchor which he loaded into his boat and carried ashore to Fish Creek.

When living on Washington Island, he was once in Ranney's store. The proprietor, to test his strength and provide some entertainment for the crowd around the store, told Bradley that he would give him a barrel of flour if he could carry it home. Without a word, Allen Bradley picked up the barrel, weighing 415 pounds, and carried it home, three miles away, without resting, and followed by a respectful and admiring crowd.

He had a long, thick beard, and it was a common amusement for some of his friends to seize it and hang suspended, whereupon Bradley would walk around the room seemingly unmindful of the burden imposed upon his chin. As the accomplishing of this feat depended on the muscles of the jaw, many doubted his ability to do it. In Green Bay a spirited wager was once made that he could not carry an ordinary man across the room in this manner. Bradley picked out Ransom Call, his old partner in the salt barrel contest. He was the heaviest man in the room, weighing two hundred and fifty pounds, and Bradley asked him to grasp his beard. This was done, and Bradley carried him with ease about the room, until Call fell down in a heap when he was no longer able to hold on.

When the Civil War broke out, he enlisted and carried his gun valiantly through many a battle. Once in a skirmish he became separated from the rest of his company and was taken prisoner by the Confederates. His musket was taken from him, and two soldiers were detailed to conduct the prisoner to the rear, while the others were ordered off on another mission. No sooner did Bradley find himself alone with his two guards, than with either hand he suddenly seized them by the neck with a grip of steel, swung them off their feet, and cracked their skulls together just once to show them what might happen if they made trouble. Then he marched them off behind the Union lines.

When the war was over and there was a general rush for the pension bureau, Allen Bradley did not join the stampede. His strength was still intact, and his sturdy independence as vigorous as ever. But after a dozen years or so old age came on. That red blood which had flowed so freely through his splendid body became chilled with his declining days and could no longer feed his massive frame. Those wonderful muscles, more tense than steel springs, became powerless and painful with rheumatism. Finally he became an object of charity, and for five years stayed at the home of a friend, Captain John Noyes, who gave him the best of care. Captain Noyes in his behalf applied for a pension, but the lords of Washington have many preoccupations. Years went by and no pension came. Finally Senator Sawyer became interested in the matter and a wretched pittance of four dollars per month was obtained.

Not until his death bed did the belated pension come.

Allen Bradley was born August 11, 1818. He came from Dunkirk, New York, to North Bay to cut cedar in 1855, but preferred to hunt and fish. He was almost impervious to cold. On the coldest days in winter he wore only a pair of trousers, socks, and moccasins, and one or two shirts. He was a striking example of the truth of the old proverb that "the race is not to the swift nor the battle to the strong." Although he had unusual opportunities, and a strength more than necessary to develop them, he never held legal title to a foot of land, and finally received a pauper's burial in the potters' field.   He died in Sturgeon Bay, February 11, 1885.

CHAPTER XV

THE RISE AND FALL OF ROWLEYS BAY

I am monarch of all I survey;
My right there is none to dispute;
From the centre all round to the sea
I am lord of the fowl and the brute.

WM. COWPER.

About a thousand miles from New York and apparently as far from a railroad, lies Rowleys Bay. It is the last little cove of Lake Michigan to the northward, dipping deep into a land of reeds and rushes, of mink and muskrat, of marsh-marigolds and fragrant balsams. At the head of the bay is a sluggish lagoon, masquerading under the name of Mink River. Here the pickerel in June are reckless and the black bass bite with abandon. Aside from these annual piscatorial activities, Rowleys Bay is as quiet and secluded as the North Pole, as indolent as the sunrise of a June morning.

But the name of Rowleys Bay has not always been the synonym of peace and pickerel. There was a time when the commercial possibilities of Rowleys Bay were eagerly discussed from Chicago to Tacoma, and glowing lithographs eloquently describing financial investments at Rowleys Bay, possible and impossible, were scattered by the tens of thousands. But we are anticipating.

Away back in the early morning of Door County's history there was a querulous old man by the name of Peter Rowley. He was one of that eccentric tribe of western pioneers who feel themselves crowded to suffocation if they have a neighbor within a day's journey. In 1836 he became oppressed by the imaginary congestion of the little frontier post at Fort Howard. He packed his possessions into a boat and fled northward past an uninhabited wilderness. Fifty miles away he came to Sturgeon Bay, as quiet and undisturbed as the morning after creation. Here at the mouth of the bay, on the west side, where now stands Cabot's Lodge, he pitched his tent, thinking he had left civilization behind forever,

But an evil fate pursued him. After a few years other eccentric pioneers followed his trail and settled in secluded coves not many miles away. On a clear day he could see the smoke from their cabin chimneys rise above the tree-tops of the distant horizon. This was intolerable. Once more he fled from congestion.

He followed the shore of Door County to its extreme northern point. Not a living soul of white men had settled north of him on the peninsula, and Peter Rowley grew hopeful. Then, as his boat was bobbing on the waves of Death's Door passage, his keen old eyes discerned the boat of a lonesome fisherman who lived at Washington Harbor, fifteen miles away. Sadly he rounded the point into Lake Michigan.

Where should he go? To the south of him lay Chicago and the pioneer camps of Milwaukee and Sheboygan. Restless fellows would soon push up the shore. In that direction lay no hope of peace. To the north was the impertinent fisherman of Washington Harbor. Where should he go? Then he discovered Rowleys Bay. He examined it carefully and believed he had discovered an oasis in the desert of civilization. Swamps to the north of him, swamps to the south of him, the great lake in front of him. Here surely was a spot where he might live and die in peace. Contentedly he reared his cabin on the shore and ate his venison and his fish. In times of extreme need he made up a raft of logs from the timber on the government land around him. In this he was assisted by two women who lived with him. Whether they were his wives, sisters, or mothers-in-law is not known. As far as we know he lived and died contentedly, his name preserved to posterity as the discoverer of Rowleys Bay.

Strictly speaking, Rowleys Bay was not discovered by Peter Rowley. A few years before he began to fish in Mink River, some other white men camped there for several weeks and ate of its fish until they loathed the sight of it. The story of this adventure is as follows:

In 1834 northern Door County was surveyed by a man named John Brink and his assistants. At one time in the fall of that year he found that provisions were running low and a messenger by the name of James McCabe was dispatched to Hamilton Arndt's trading post at Green Bay for supplies. Mounted on a trusty pony, named Polly, the messenger started off with instructions to join Brink and his men at a certain place near Death's Door in three weeks.

The trip to the Indian trader's was made without incident, but on his return, when not far from Death's Door, he was taken prisoner by a band of Indians, who thought he was a deserter from the army. McCabe was about one hundred yards from the pack horse at the time, having stopped in a grove to camp over night. When the Indians seized him they did not know that he had a horse with him, and they would not, or rather could not, let him explain, as he did not understand their speech.

The Indians were sometimes called upon to assist the soldiers in running down deserters, and when they were of any assistance they were always supplied with a little whiskey for their services. With the prospect of getting some "fire-water" for the return of McCabe to the government fort, they watched him carefully. The more he remonstrated the more the Indians believed he was a deserter.

McCabe, therefore, not knowing whether the red men intended to burn him at the stake, was compelled to go with the Indians, while Polly, with the pack of provisions, was left grazing in the little grove.

Those fool Injuns actually made McCabe carry a canoe five miles across the peninsula, said Mr. Brink when telling the story, and he was taken to Hamilton Arndt's headquarters, where the Injun trader had some difficulty in making the varmints believe that McCabe was not a deserter from the army.

All this time, we, of course, were waiting for the packman at the place appointed, and were without anything to eat, having waited two days, and lived during that time on nothing but hope. Still no packman, and we had no firearms to kill game, even if any could have been found. At the expiration of two days you can imagine that we were pretty hungry. We concluded to get something to eat when the third day rolled around, and we moved on toward the lake and discovered a little creek running into it.

As luck would have it the stream was full of fish, and we had no trouble in catching all the big fellows we wanted. There was one man in our party who was so hungry that he didn't even wait to cook the fish. He just scraped off the scales and chewed the stuff up almost before the finny creature was dead.

For just eleven days we lived on nothing but roasted fish. It was fish for breakfast, fish for dinner, and fish for supper, and you had better believe we were sick of fish before we got through with our experience. We had no salt or anything to flavor the stuff with. It was simply roast fish day after day. It sickened me of fish and I haven't eaten any since. It kept life in us, however. When relief did come it came unexpectedly.

The twelfth day, when we arose to begin the day with a fish breakfast, we heard the tinkling of a bell, and on the crest of a little hill we saw old Polly. As soon as she discovered us she came galloping up, neighing as if overjoyed to see us. She was so pleased to see us that she actually laughed. I could see her eyes blaze with delight, and as she rubbed her nose against my shoulder she appeared to be brimful of happiness.

The pack containing the pork and beans and flour was still strapped to her back, and you can bet all you have got that we had a good square meal that day. As far as we could learn, Polly had gone back to the place from where McCabe had started with her, and, not finding us there, had wandered around the country following our trail, and finally discovered us.

The next day McCabe appeared, having been released as soon as the Indian trader explained matters to his captors. He expected to find a rather sickly looking lot of men, and if he didn't find what he thought he would, he certainly did find a fishy crowd, for we were covered with scales and smelled like the inside of a whale.

The history of Rowleys Bay for the next thirty or forty years is a blank as far as human interest is concerned. Gradually the lumber companies found their way thither. Camps were built where the men sat in their bunks and swapped stories of the woods. A pier was built, and huge cargoes of telegraph poles, ties, and cordwood were shipped. The work of destruction pursued the even tenor of its way.

In 1876 S. A. Rogers arrived from New York. He had a farm in Illinois which, through the medium of a real estate agent, he traded off for a vast acreage of land and water at Rowleys Bay. Unfortunately, the land and water were mixed together after a somewhat haphazard formula, constituting a 4000-acre tract of swamp land covered with a pretty good stand of cedar. Being a man of energy, Mr. Rogers built a large sawmill which sometimes scaled a run of seven or eight million feet of lumber in a season. He built a commodious pier along which nearly always lay a vessel or two loading. He also built a store and other buildings for the accommodation of the growing business of the place.

All this business centered in the cedars which were large enough to cut. But there were millions of cedars too small even to make a fence post. Of what use were they? Much cogitation on this subject followed.

About 1885 a man was found who solved this puzzle. This was J. H. Mathews of Milwaukee, who understood the process of making cedar oil. He built a factory on the northeast side of Rowleys Bay, where he employed about twenty-five men. Cedar twigs were cut and placed in a tank or retort. The dimensions of this retort were four by twenty-two by eight feet, the top being convex. The steam from this retort was taken up into a four-inch pipe and cooled and conducted through a succession of pipes of decreasing diameters placed zigzag fashion in a bed of a small creek fed by cold spring water. After the steam had meandered through these cold pipes for a distance of about two hundred feet, it trickled into a receiving tank in the shape of limpid oil which sold at eight dollars per gallon. For two years the business was pushed and paid very well.

Mr. Mathews was a man of enterprise and ambition. He reasoned that if good money could be made out of waste timber products in such an inaccessible place as Rowleys Bay, much more could be made if the business was enlarged and established in a more central place. Accordingly, he pulled up his cooling pipes and moved to Marshfield, Wis., where he undertook to make wood alcohol. He promptly failed in business, and with this his part in the history of Door County is finished.

About 1892 Mr. Rogers found an opportunity for trading off several hundred acres of his swampy

estate for a farm in Missouri. Through another
trade this tract of swamp land was transferred to
a Mr. Ditlef C. Hanson, a thrifty little Dane of
Tacoma, Washington. In the course of time
Mr. Hanson came to inspect his purchase.

He found the land too low for farming, too high
for fishing. The timber was all gone. It was too
inaccessible for a frog preserve, and muck was
drug on the market. What was it good for?

Mr. Hanson had one great ambition in life. He
had heard of other men laying out a townsite,
waxing rich by the sale of building lots, and
famous by having the town named after them.
He reasoned that since his Rowleys Bay possession
was fit for nothing else, if it was not created in
vain it must have been intended for a townsite.
True, it was wet, but Mr. Hanson, being a man of
reading, recalled that a wet foundation was no
barrier to the most shining successes in city
building. Chicago was built in a marsh. Venice
was built in a lagoon, and Shanghai was originally
a frog pond. A townsite then it was to be, forever
to immortalize its founder, Ditlef C. Hanson. He
debated whether to call it "Ditlef's Hope" or
"Hansonburg," but finally rejected both as lacking
in euphony. Instead, he named it Tacoma Beach,
which was both resonant and reminiscent of the
city of his home. This important point being
settled, he immediately sought a printer.

Townsite lithographs are wonderful things.
In 1836 a city was platted about where is now the
present city of Kewaunee, a lake port some
distance south of Rowleys Bay, and large fortunes

were made and lost by means of an eloquent lithograph. A nomadic fur-trader had shortly before picked up something in the swamp at the mouth of Kewaunee River, which his imagination had transmuted into gold. Rumor reached the ears of some enterprising promoters who proceeded to lay out a townsite. Not a settler at that time lived within thirty miles of the place, but that did not prevent the project from becoming a great transient success. A number of men of national fame became interested, among them being such men as John Jacob Astor, Governor Doty, Governor Beals, Judge Morgan L. Martin, Hon. Sanford E. Church, General Ruggles, Colonel Crocker and Salmon P. Chase, later Chief Justice of the Supreme Court of the United States. For a while there was much debate in the minds of great financiers whether to invest in Chicago or Kewaunee real estate. In April, 1836, a forty-acre tract in the swamp was sold to Governor Doty for $15,000. Judge Martin had entered a tract of eighty acres in the swamp from the government. This he sold within a few days to his distinguished colleague, Chief Justice Chase, for $38,000. These and other lands were subdivided into lots and on September 2, 1836, a grand auction was held in Chicago. There was a great rush for the lots, some selling as high as a thousand dollars, and the promoters reaped barrels of money. For a while there was much slushing around in top boots in the Kewaunee swamp in search of gold. Nothing was found, the investors went sadly away,

and the land reverted into an untaxed and unsettled wilderness for the next thirty years.

Our Ditlef C. Hanson had no such rosy dreams
of success. He did not know any governors or
supreme court justices. But he did his best with
the material in hand. He got out a stock of
splendid lithographs. These showed a townsite
plat more than a mile long with wide streets and
curving avenues. No such common names as
Pike Street or Billings Avenue were here permitted. They were all sonorous street names,
reminiscent of the glory of the republic, such as
Arlington Avenue, Columbia Street, Potomac
Boulevard, etc. Along the shore a beautiful park
was shown, enlivened by smart carriages and gay
children dashing around on roller skates. Some
streets were marked with street car lines, and
certain corners were marked as occupied by a
public library, postoffice, sanitarium, bank, or
other institutions of importance. Even sluggish
old Mink River, as if taking new life by this
activity, was pictured as a dashing stream, leaping
over boulders and plunging at last into the lake
by means of an inspiring waterfall. All in all it
was the most imposing document ever published
setting forth the charms of Door County.

Armed with these lithographs, Mr. Hanson
returned to Tacoma and opened the campaign.
He showed them to friends and foes, who were
duly impressed and sometimes bought. He discovered, however, that the vastness of the American continent lying between Tacoma and Tacoma

Beach deterred many who would otherwise have eagerly invested. Because of this, and because, like Moses, he was slow of speech, though of great resource, he determined to go to Chicago and sell out. He went to Chicago where he met a man with a name something like Rosenstein. To him he sold his entire stock of lithographs, with the townsite thrown in.

Mr. Rosenstein was enthusiastic about his purchase. He went out into the highways and byways of the city and explained the lithographs to all who would listen. He showed them how they could live happily at Tacoma Beach, or, if not, how they could die, secure in the faith that their money was well invested and that their widows would bless their memory. His arguments were irrefutable.

In due course of time many of these investors came to view the paradise of their purchase. Among them was a semi-invalid who came with a full equipment of paints, pots and brushes. He had taken the job of painting the cottages of the new city. Some went as far as Sturgeon Bay, others went on to Fish Creek and Sister Bay, while still others persisted in pushing on to Rowleys Bay, before they were disillusioned. Alas, they each and all discovered that they had forgotten the most important part of their equipment for viewing the new city—top boots.

We will not linger over the gnashing of teeth or the bitter recriminations heaped upon old Rosenstein. The lots were sold and the lithographs used up, so he merely shrugged his shoulders and turned his thoughts to other things. So, also, after a while, did the dupes. Their money was gone, so they wasted no more in paying taxes on their submerged lots on Lakeside Boulevard. It remained now merely for the long suffering county board to unravel the tangle. Finally the "streets" were vacated and the land sold for taxes. The affair cost the county about five thousand dollars.

After eight years' flight in financial circles Rowleys Bay returned once more to its undisturbed seclusion. In the parks of the new city the frogs croak by day and the crickets chirp by night. Even frisky old Mink River has ceased from its gambols, and settled into its sluggish solitude where the pickerel in June are reckless and the black bass bite with abandon.

CHAPTER XVI

TOILERS OF THE SEA

There dwells a wife by the Northern Gate,
And a wealthy wife is she;
She breeds a breed of roving men
And casts them over sea,
And some are drowned in deep water,
And some in sight o' shore.
And word goes back to the weary wife,
And ever she sends more.

KIPLING.

Door County has a shore line of about two hundred miles, not counting the innumerable smaller bays and inlets which indent its shores. This shore line is faced with reefs and headlands and along its front are scores of hidden shoals and dangerous passages. For many years quite as much of Door County's history was enacted on the water as on the land. The many hundred fishermen daily watched the vagaries of the sea, and the alert sea captains, traveling through these waters with their cargoes of forest products, learned to respect the sudden gales of Green Bay, the treacherous squalls in "the Door" and the big storms of Lake Michigan. The passage through "the Door" was particularly dreaded. To lose a deckload in making the passage was so ordinary an event as scarcely to be worthy of mention. Before the writer lies the diary, or log-book, of

232

the lighthouse keeper of Pilot Island, kept from 1872 to 1889. It appears from the entries in this journal that their daily experience consisted of winds and roaring seas with a shipwreck at least twice a week. It seemed the regular thing in those days of sailing vessels to ground on a shoal, throw off the deckload, and then work loose. Frequently the keeper or one of his assistants, on his occasional mail trips to Washington Island, is unable to return for three weeks. Owing to the extremely exposed position of this little rock in the sea, it is practically inaccessible in any storm. In the fall of 1872 he reports eight large vessels stranded or shipwrecked in "the Door" in one week. The preceding year, 1871, almost a hundred vessels were lost or seriously damaged in passing through "the Door."

The greatest storm of present memory occurred October 16, 1880. It started to blow from the southeast on the evening of October 15th, and continued for three days. The waves ran so high that at the Cana Island Lighthouse the sea frequently broke over the house. The lantern at a height of eighty-eight feet was at times completely covered with spray from the huge waves.

This storm did immense damage to shipping. At "the Door" twelve vessels were driven on the rocky beaches of Plum Island and Detroit Island and were seriously damaged, many of them being a total loss. At Baileys Harbor seven large vessels were stranded, two being a total loss. In North Bay a large fleet sought refuge from the storm.

The Crags of North Bay

The vessels were mostly of the larger class and were either bound for Buffalo with grain or were on their way from that port to Chicago with coal and other supplies. About fifty of these vessels crowded into the little harbor during the storm. Some of them made the harbor safely and anchored. As the storm increased in fury some of these dragged their anchors and were driven ashore. Here the waves washed over them and it was necessary to throw out the cargoes to keep the decks from bursting from the swelling grain. As more and more vessels came in there were a number of collisions, several large vessels sank, and the confusion and distress was indescribable. The crew from one of these vessels was rescued in such a daring manner by one of the fishermen on shore that mention must here be made of it.

At five o'clock in the afternoon a large schooner named *Two Friends* attempted to enter the narrow inlet. A part of her canvas was blown away so she could not be kept on her course, but drifted on the rocky northern point where she struck the limestone ledge in twelve feet of water. Here she lay exposed to all the fury of the storm. The first wave that swept over her after she grounded swept the yawl from its davits and carried it away. The crew attached lines to fenders hoping they would drift ashore, but the current was so strong past the point that the plan failed. The tremendous pounding of the waves soon broke up the deck and carried the main and mizzen-masts overboard, whereupon the crew of seven men took refuge in the forward rigging which still held firm though groaning ominously. Their shouts for help were plainly heard on shore but the sea was running so high that help seemed impossible. The group of fishermen on the shore discussed the possibility of rescue, but gloomily agreed that nothing could be done. There remained only the morbid curiosity to see the boat soon go to pieces and seven wretches perish before their eyes.

Then appeared a young Danish fisherman by the name of James Larson. After taking in the situation he approached the fishermen and vehemently urged that an attempt be made to save the crew on the vessel. But one and all emphatically shook their heads. How could a boat be launched in such a surf? And even if it could be, how would it force its way through such immense breakers? Such an attempt was not only foolhardy, it was suicidal.

But Larson was undismayed. If he could get no help from the others, he determined to attempt the rescue alone. Out there were seven human beings who would surely perish unless help was brought to them. To attempt it very likely meant the loss of his own life, but it was the chance of risking one life against seven. The odds were worth it.

He obtained a light line long enough to reach from the shore to the schooner which lay about six hundred feet off shore. It was his plan to carry this line out to the wreck and with this to pull a heavy hawser back to shore by help of which the shipwrecked sailors could worm their way to safety.

Being alone to row the boat, he selected the smallest and lightest skiff on shore. The owner strenuously objected as he felt confident it meant the loss of the boat, but Larson took it by force. Then with the help of two men to push the boat through the surf he jumped in and seized the oars. It was now about eleven o'clock, but the moon was giving a moderate amount of light.

For a while it seemed impossible to make any headway against the surging, white-crested waves, but with intense exertions Larson stuck to his task, and little by little forged ahead. At last he reached the lee of the schooner, but found that the sailors were so chilled that they were incapable of going ashore by help of a hawser as he had planned. The only possible alternative was to take them ashore in the boat, but this was so small that only one man could be taken at a time and the passenger had to lie flat on his back to avoid capsizing. Larson was therefore obliged to make a separate trip for each sailor. Six times the boat was filled with water and he had to return to the shore to make a fresh start. But Larson indomitably continued his heroic task until the last man was safely landed.

In order to minimize the dangers that lurk around Door County's rock-bound shore the government has erected seventeen lighthouses. On islands, hills and promontories they rear their whitewashed towers, their lights blinking through the night to guide the storm-tossed vessels to safety.

The first to be erected was the "Pottawatomi Light." This was built in 1836 and was the first lighthouse on Lake Michigan. It was built on the northwesterly point of Rock Island as the most used channel of navigation, before Chicago and Milwaukee had gained prominence, lay north of Washington Island. It is the highest lighthouse on all the inland lakes, its lantern being 137 feet above sea level. Its light can be seen for twenty miles.

The next lighthouse to be erected was the Pilot Island Light, known as the Port des Morts Station. It was not until about 1850 that "the Door" passage between Washington Island and the peninsula was much used for navigation and the light was then erected to guide the mariners through this dangerous passage. This lighthouse is frequently enveloped in fogs which settle there. At such times its ten-inch steam whistle roars out its powerful warning every thirty seconds. Pilot Island is a mere rock in the sea and a dreary place to stay in. Victor E. Rohn, a lieutenant in the Civil war, who was chief lighthouse keeper from August, 1872, to November, 1876, makes the following observation under date of July 4, 1874:

Independence Day came in fine after a heavy southeast gale. This island affords about as much independence and liberty as Libby Prison, with the difference in guards in favor of this place, and chance for outside communication in favor of the other.

The following description of Pilot Island gives a pleasanter picture of life there. It was written by a visitor to the island in 1890.

Pilot Island is a little island of three and a quarter acres of rock and boulders on which there is an imported croquet ground, a few ornamental trees, a strawberry patch, two fog sirens, a lighthouse, a frame barn, a boat house and some blue, bell-shaped flowers and golden-rods that grow out of the niches in the rock. It lies about two miles east of Plum Island, which is so called because it lies plumb in the center of Death's Door.

This is truly an isolated spot, but I have spent five days on Pilot Island and they are among the happiest days of my experience. At sunrise every morning the first assistant, Chas. E. Young, would wake us up with an invitation to go in bathing. Then the keeper and second assistant and I would leave our cozy beds, run down with him to the landing and plunge into the almost ice cold water. We would swim to the leeward of the island where the breakers meet as they come from both sides. Every few minutes they would crash together and hoist us into the air in the midst of a cloud of foam and spray.

On moonlight nights it is like being in a dream of ideality to walk alone over the moss-covered rocks and listen to the swish of the surf as it charges over the breakwater at the boatlanding, hear it roaring on all sides of the little island and see huge vessels under full sails crossing the moon-glade on their way through "the Door." One seems to be completely separated from all that is worldly and bad. There is no field for gossip here. The land is not suitable for general farming purposes, but it is a splendid place to raise an ample crop of good, pure thoughts.

One of the fog sirens at this station is an exact duplicate of the one that was on exhibition at the World's Fair held in Paris. Its song can be heard a distance of forty miles, and when it sings all the lights in the signal house must be hung by strings to prevent them from going out. The sound is so tremendous that no chickens can be hatched on the island, as the vibration kills them in the egg and it causes milk to curdle in a few minutes. Visitors at the lighthouse on foggy nights sit up in bed when the siren begins its lay and look around for their resurrection robes.

The Pilot Island lighthouse is famous for having witnessed more shipwrecks than any other lighthouse on the Great Lakes. If their number could be told it would be a legion. On this little crag and its near by rocks and shoals scores of proud vessels have been irresistibly driven to be quickly pounded to pieces by the thundering seas. Many times the keepers of the lighthouse have been called upon at the risk of their lives to save the imperilled crews. A notable example of this was the heroic rescue of the *A. P. Nichols* on November 9th, 1892, by the keeper of the light, Martin Knudson.

The Nichols was bound from Chicago to Escanaba without cargo and was caught in a fearful storm. Her big anchor was no match for the gale and the schooner drifted on the rocks of Pilot Island. When she struck the waves washed clean over her.

Martin Knudson, the keeper of the light, was familiar with every rock around his storm-beaten island and knew the location of a shoal leading out to the stranded vessel. It was late at night and intensely dark, and the waves came crashing in with terrific force over the slippery rocks. In spite of these handicaps he waded out along the shoal, up to his neck in water until he almost reached the vessel. With much shouting he finally made himself heard above the roar of the waves and ordered the captain and his men to jump, one by one, and he would catch them. It seemed like suicide to jump into that foaming cauldron, but in order to see if rescue was possible Captain Clow jumped first. He went in far over his head, but Knudson caught him before he was sucked away by the undertow. The captain remained on the shoal while Knudson rescued the next one in the same way. In this manner the crew of six was rescued, including a female cook and the captain's aged father, old Captain David Clow, who had suffered shipwreck at nearly the same spot twenty years before. When the last one was caught Knudson piloted them all ashore along the narrow and crooked ridge of the shoal.

Speaking of this rescue Captain Clow said later: "It is a wonder to me how Knudson found his way along that ledge of rocks in the darkness of the night. He is about the bravest man I have ever seen. How he managed to keep his bearings after rescuing the crew, has been a wonder to me ever since. A single misstep and we would all have fallen off the rocks into deep water and undoubtedly been drowned."

The Schooner *J. C. Gilmore* had gone to pieces on the island a week before and her crew was still at the lighthouse. The addition of the crew of *the Nichols* made sixteen persons to feed, and for a while it looked to the men as if they had escaped drowning only to die of starvation. For a week the storms prevented any one from leaving the island.

However, a little lull enabled the lighthouse crew to obtain some provisions from *the Nichols,* which soon afterward went to pieces.

Besides the many lighthouses there are also a number of gas buoys whose constant glimmer bids the mariners to take heed of the dangerous shoals they mark. One of these is the notorious whaleback shoal lying about six miles west of Washington Island.

Whaleback shoal is about two miles long and lies nearly in the direct line of travel up and down Green Bay. It extends in a straight line from southeast to northwest and craggy portions of its narrow ridge rise above the surface of the water. On quiet days the water flows placidly over this submerged ridge and the surface gives no indication of the danger that is lurking below. In stormy weather, however, the sea froths and roars over it in a terrific manner. In the early days, before the bay was properly charted and danger spots marked, this submerged reef was the cause of many a marine disaster.

To the credit of Whaleback Shoal it must be told that once it was the means of saving lives instead of destroying them. Two fishermen from Ephraim by the names of Anton Olson and Anton Amundson were fishing on the ice near Chambers Island early in the spring of 1890. While they were busily engaged in pulling up one trout after another the ice broke up and the fishermen found themselves afloat on a large floe. It began to break up and they knew that sure death awaited them if they were carried out into Lake Michigan. After spending a night and a day of fading hopes and weariness on their crumbling raft they found themselves close to Whaleback Shoal. Covered with large cakes of ice which had stranded there, it now lay like a huge, sinister serpent of ice. It had one virtue, however, it was not moving toward sure death. With their ice picks the fishermen broke loose a small cake of ice and by means of this ferry reached the shoal. Here they found the ice heaped up in the greatest confusion, making strange caves and crevices. They crawled into one of these though with little hope. They were too far from land to be seen and they knew that no vessel would venture out for several weeks. Moreover, the shoal was such a dangerous place that sailors gave it a wide berth.

For two days and nights they sat in their cages of ice, exposed to the freezing cold of early April nights without food. The third morning the wind shifted and a mass of broken ice drifted southward. Finally, some cakes came close to the shoal and the fishermen jumped for them. By making hair-breadth leaps from one small chunk to another they thus made their way for twelve miles over the slippery, bobbing ice and at last reached land north of Sister Bay.

Door County may be said to be a nursery of lighthouse keepers and coast guards. It has only three coast guard stations of its own, but sons of Door County are to be found in the crews of the coast guard stations on all the lakes and the

Atlantic seaboard. Ninety percent of the commanders of all the life saving stations on the west shore of Lake Michigan hail from the peninsula. No other county in America has so many representatives in the intrepid business of watching the thousands of miles of coastal waters of the United States.

It is also preëminently the home of the sailor. Door County boys are nearly all the sons of fishermen or of sailors turned farmers, and with their constant childhood vision of the sea they turn as naturally to a seafaring life as the Welsh turn to mining. On all sorts of crafts they are to be found, from the huge carferries of the Great Lakes crashing their course through three feet of ice with thirty loaded freight cars in the hold, to the mammoth ocean liners that ply between foreign ports. But whether in Santiago or Singapore the Door County sailor looks back to his native home with its headlands and variously indented shores, its forests and green fields, and its glorious sunsets, as the fairest spot on earth.

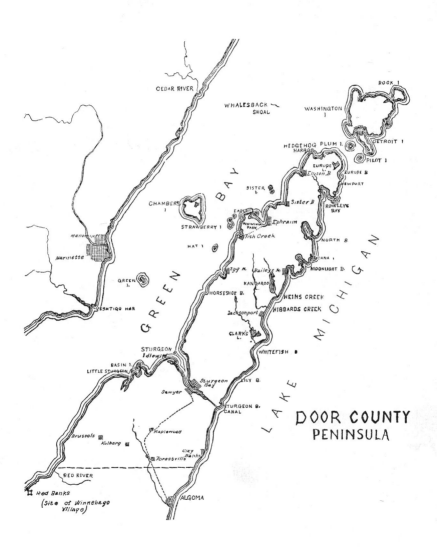

CEDAR RIVER

WHALESBACK
SHOAL

WASHINGTON
I.

ROCK I.

HEDGEHOG PLUM I.
HARBOR

DETROIT I.

PILOT I.

EUROPE I.
Ellison B.

EUROPE B.

NEWPORT

CHAMBERS
I.

SISTER
I.

Sister B

ROWLEY'S
BAY

STRAWBERRY I.

EAGLE

Peninsula
PARK

Ephraim

MENOMINEE

Fish Creek

NORTH B.

HAT I.

MARINETTE

Egg H.

Bailey's H.

CANA I.

MOONLIGHT B.

GREEN
I.

KANGAROO

HORSESHOE B.

HEINS CREEK

PESHTIGO HAR.

Jacksonport

HIBBARDS CREEK

CLARK'S
L.

STURGEON
Idlewild

WHITEFISH B.

BASIN I.
LITTLE STURGEON

Sturgeon
Bay

LILY B.

Sawyer

STURGEON B.
CANAL

Brussels

Maplewood

Kulberg

Clay
Banks

Forestville

RED RIVER

Red Banks
(Site of Winnebago
Village)

ALGOMA

GREEN BAY

LAKE MICHIGAN

DOOR COUNTY
PENINSULA